BIONICS: ARTIFICIAL LIFE PARTS

BIONICS: ARTIFICIAL LIFE PARTS

Edited by:

Shiv Sanjeevi and Dr Prerna Pandey

Delve
PUBLISHING

www.delvepublishing.com

Bionics: Artificial Life Parts

Shiv Sanjeevi and Dr Prerna Pandey

Delve Publishing
2010 Winston Park Drive,
2nd Floor
Oakville, ON L6H 5R7
Canada
www.delvepublishing.com
Tel: 001-289-291-7705
001-905-616-2116
Fax: 001-289-291-7601
Email: orders@arclereducation.com

ISBN: 978-1-77361-186-0 (Hardcover)

DECLARATION

Some content or chapters in this book are open access copyright free published research work, which is published under Creative Commons License and are indicated with the citation. We are thankful to the publishers and authors of the content and chapters as without them this book wouldn't have been possible.

ABOUT THE EDITORS

Shiv Sanjeevi Sripathi completed his BSc in Biotechnology from Mumbai University and secured second rank in Mumbai University and first rank in his college: Kishinchand chellaram College. Shiv secured first rank in his college KET's V.G.Vaze College and has worked on a stem cell project at the Specialized Centre for Cell Based Therapy (SCCT), KEM Hospital at Mumbai, following which he was a Junior Research Fellow at Microbiology & Cell Biology Department, Indian Institute of Science, Bangalore from August 2009 to November 2011. The field of work involved cloning of cell wall genes and transcription factors in E. coli & M. smegmatis. This was followed by a stint as a Research and Development Executive at Transasia Biomedicals, Mumbai where he worked on patent writing and developing immunoassays from 2011 till 2013. In 2013, he entered the teaching field where he has been tutoring biology and chemistry to the students of 11th and 12th grades along with English grammar and communication skills to people of all age groups.

Dr Prerna Pandey, a biotechnologist with several years of wet lab research experience, worked at International Center for Genetic Engineering and Biotechnology, New Delhi. She has also worked at Transasia Biomedicals and Advance Enzyme Technologies as a scientist. Prerna has published papers in peer-reviewed journals, and has submitted a number of annotated Geminiviral genome sequences in the GenBank, including two novel ones. Now, she works as a freelance scientific editor and writer.

TABLE OF CONTENTS

LIST OF CONTRIBUTORS

Shiv Sanjeevi Sripathi
Shiv Sanjeevi Sripathi completed his BSc in Biotechnology from Mumbai University and secured second rank in Mumbai University and first rank in his college: Kishinchand chellaram College. Shiv secured first rank in his college KET's V.G.Vaze College and has worked on a stem cell project at the Specialized Centre for Cell Based Therapy (SCCT), KEM Hospital at Mumbai, following which he was a Junior Research Fellow at Microbiology & Cell Biology Department, Indian Institute of Science, Bangalore from August 2009 to November 2011. The field of work involved cloning of cell wall genes and transcription factors in E. coli & M. smegmatis. This was followed by a stint as a Research and Development Executive at Transasia Biomedicals, Mumbai where he worked on patent writing and developing immunoassays from 2011 till 2013. In 2013, he entered the teaching field where he has been tutoring biology and chemistry to the students of 11th and 12th grades along with English grammar and communication skills to people of all age groups.

Prerna Pandey
Dr Prerna Pandey, a biotechnologist with several years of wet lab research experience, worked at International Center for Genetic Engineering and Biotechnology, New Delhi. Her field of research in PhD included isolation and molecular characterization of geminiviruses, genome sequencing, gene annotation, and gene silencing using the RNA interference technology. She has also worked at Transasia Biomedicals and Advance Enzyme Technologies as a scientist. Prerna has published papers in peer-reviewed journals, and has submitted a number of annotated Geminiviral genome sequences in the GenBank, including two novel ones. She has also completed her editing and proof-reading courses from Society for the promotion of Editors and Proof-readers, UK and now works as a freelance scientific editor and writer. When Prerna is not busy with her assignments, she loves to explore historical places.

Elliott J. Rouse
Biomechatronics Group, MIT Media Lab, Massachusetts Institute of Technology, Cambridge, Massachusetts, United States of America

Nathan C. Villagaray-Carski
Biomechatronics Group, MIT Media Lab, Massachusetts Institute of Technology, Cambridge, Massachusetts, United States of America
Department of Electrical Engineering and Computer Science, Massachusetts Institute of Technology, Cambridge, Massachusetts, United States of America

Robert W. Emerson
A Step Ahead Prosthetics, Burlington, Massachusetts, United States of America

Hugh M. Herr
Biomechatronics Group, MIT Media Lab, Massachusetts Institute of Technology, Cambridge, Massachusetts, United States of America

Lauren N. Ayton
Centre for Eye Research Australia, University of Melbourne, Royal Victorian Eye and Ear Hospital, East Melbourne, Australia

Peter J. Blamey
Bionics Institute, East Melbourne, Australia
Department of Medical Bionics, University of Melbourne, East Melbourne, Australia

Robyn H. Guymer
Centre for Eye Research Australia, University of Melbourne, Royal Victorian Eye and Ear Hospital, East Melbourne, Australia

Chi D. Luu
Centre for Eye Research Australia, University of Melbourne, Royal Victorian Eye and Ear Hospital, East Melbourne, Australia

David A. X. Nayagam
Bionics Institute, East Melbourne, Australia
Department of Pathology, University of Melbourne, St Vincent's Hospital Melbourne, Fitzroy, Australia

Nicholas C. Sinclair
Bionics Institute, East Melbourne, Australia

Mohit N. Shivdasani
Bionics Institute, East Melbourne, Australia
Department of Medical Bionics, University of Melbourne, East Melbourne, Australia

Jonathan Yeoh
Centre for Eye Research Australia, University of Melbourne, Royal Victorian Eye and Ear Hospital, East Melbourne, Australia

Mark F. McCombe
Centre for Eye Research Australia, University of Melbourne, Royal Victorian Eye and Ear Hospital, East Melbourne, Australia

Robert J. Briggs
Centre for Eye Research Australia, University of Melbourne, Royal Victorian Eye and Ear Hospital, East Melbourne, Australia

Nicholas L. Opie
Centre for Eye Research Australia, University of Melbourne, Royal Victorian Eye and Ear Hospital, East Melbourne, Australia

Joel Villalobos
Bionics Institute, East Melbourne, Australia

Peter N. Dimitrov
Centre for Eye Research Australia, University of Melbourne, Royal Victorian Eye and Ear Hospital, East Melbourne, Australia

Mary Varsamidis
Centre for Eye Research Australia, University of Melbourne, Royal Victorian Eye and Ear Hospital, East Melbourne, Australia

Matthew A. Petoe
Bionics Institute, East Melbourne, Australia

Chris D. McCarthy
NICTA, Computer Vision Research Group, Canberra, Australia
National Institute for Mental Health Research, Australian National University, Canberra, Australia

Janine G. Walker
NICTA, Computer Vision Research Group, Canberra, Australia
National Institute for Mental Health Research, Australian National University, Canberra, Australia

Nick Barnes
NICTA, Computer Vision Research Group, Canberra, Australia
National Institute for Mental Health Research, Australian National University, Canberra, Australia

Anthony N. Burkitt
Bionics Institute, East Melbourne, Australia
Centre for Neural Engineering, University of Melbourne, National Information and Communications Technology Australia (NICTA), Ltd., Melbourne, Australia

Chris E. Williams
Bionics Institute, East Melbourne, Australia

Robert K. Shepherd
Bionics Institute, East Melbourne, Australia
Department of Medical Bionics, University of Melbourne, East Melbourne, Australia

Penelope J. Allen
Centre for Eye Research Australia, University of Melbourne, Royal Victorian Eye and Ear Hospital, East Melbourne, Australia

Philippe Malcolm
Department of Movement and Sports Sciences, Ghent University, Ghent, Belgium

Wim Derave
Department of Movement and Sports Sciences, Ghent University, Ghent, Belgium

Samuel Galle
Department of Movement and Sports Sciences, Ghent University, Ghent, Belgium

Dirk De Clercq
Department of Movement and Sports Sciences, Ghent University, Ghent, Belgium

Yiyun Zhu
School of Civil Engineering and Architecture, Xian University of Technology, No. 5 Jinhua South Road, Xian, Shaanxi 710048, China

Chang Jiang Wang
School of Engineering and Informatics, University of Sussex, Brighton BN1 9QJ, UK

Diane Mynors
School of Engineering and Informatics, University of Sussex, Brighton BN1 9QJ, UK

Ulrich Lüttge
Department of Biology, Technical University of Darmstadt, Schnittspahnstr. 3-5, 64287 Darmstadt, Germany

Barbara Mazzolai
Center for Micro-BioRobotics, Istituto Italiano di Tecnologia, Pontedera, Italy

Lucia Beccai
Center for Micro-BioRobotics, Istituto Italiano di Tecnologia, Pontedera, Italy

Virgilio Mattoli
Center for Micro-BioRobotics, Istituto Italiano di Tecnologia, Pontedera, Italy

Michael A. McCloskey
Department of GDCB, Iowa State University, Ames, IA, United States of America
Creodyne, LLC, 2410 NE 13th Ct, Ankeny, IA, United States of America

Curtis L. Mosher
Department of GDCB, Iowa State University, Ames, IA, United States of America

Eric R. Henderson
Department of GDCB, Iowa State University, Ames, IA, United States of America
Creodyne, LLC, 2410 NE 13th Ct, Ankeny, IA, United States of America

PREFACE

The potential of connecting technology between live beings and manufactured items opens up a vista of exciting research. The topic of "bionics" explores the study of mechanical systems that function like living organisms or specific parts of living organisms: that seems straight from scenes of a science fiction tale. But, with each passing day, the connection between technology and biology seems to open up several interesting and sometimes a boon to mankind.

The field of bionics encompasses the construction and use of robotic arms that offer relief to the maimed, electrodes reroute nerves to achieve specific pathways, tissue engineering, cochlear implants, artificial hearts to bionic eyes to name a few. Bionics research has also propelled research in the field of aviation and computer science (an interesting field of artificial neural networks).

This book shall include the history, research into the field of bionics along with various aspects of the products/technologies that use bionics and the scope of this field of research. Through this book, the editors hope to increase the interest and research in this field to open up vistas in research.

Editors

CHAPTER 1

BIONICS: AN INTRODUCTION AND CONCEPTS

Shiv Sanjeevi Sripathi[1] and Prerna Pandey[2]

[1]Shiv Sanjeevi Sripathi completed his BSc in Biotechnology from Mumbai University and secured second rank in Mumbai University and first rank in his college: Kishinchand chellaram College. Shiv secured first rank in his college KET's V.G.Vaze College and has worked on a stem cell project at the Specialized Centre for Cell Based Therapy (SCCT), KEM Hospital at Mumbai, following which he was a Junior Research Fellow at Microbiology & Cell Biology Department, Indian Institute of Science, Bangalore from August 2009 to November 2011. The field of work involved cloning of cell wall genes and transcription factors in E. coli & M. smegmatis. This was followed by a stint as a Research and Development Executive at Transasia Biomedicals, Mumbai where he worked on patent writing and developing immunoassays from 2011 till 2013. In 2013, he entered the teaching field where he has been tutoring biology and chemistry to the students of 11th and 12th grades along with English grammar and communication skills to people of all age groups.

[2]Dr Prerna Pandey, a biotechnologist with several years of wet lab research experience, worked at International Center for Genetic Engineering and Biotechnology, New Delhi. Her field of research in PhD included isolation and molecular characterization of geminiviruses, genome sequencing, gene annotation, and gene silencing using the RNA interference technology. She has also worked at Transasia Biomedicals and Advance Enzyme Technologies as a scientist. Prerna has published papers in peer-reviewed journals, and has submitted a number of annotated Geminiviral genome sequences in the GenBank, including two novel ones. She has also completed her editing and proof-

reading courses from Society for the promotion of Editors and Proof-readers, UK and now works as a freelance scientific editor and writer. When Prerna is not busy with her assignments, she loves to explore historical places.

FEW DEFINITIONS OF BIONICS

When pressed with a technical problem, humans regularly draw guidance and notion from the herbal world. Through the process of evolution, organisms have experimented with form and feature for at least three billion years earlier than the first human manipulations of stone, bone, and other materials. Although we cannot know for sure the volume to which organic models stimulated our early ancestors, more current examples of designs are properly documented. For instance, birds and bats performed a vital function in one of the more triumphant feats of human engineering, the construction of an aircraft.

Though the origin of bionics may be traced back to Leonardo da Vinci whose mechanical drawings were inspired by many elements of nature, the tale of Icarus who flew to the Sun using wings has been well documented.

The term *bionics* was coined by Jack Steele of the US Air Force in 1960 at a Wright-Patterson Air Force Base meeting at in Ohio. The definition is explained as a function that mimics nature, or represents features of natural systems or their analogues. It was reported that Leonardo da Vinci studied birds in flight and designed machines (Vincent JF et al., 2006).

Bionics is the application of biological methods and systems found in nature to the study and design of engineering systems and modern technology (Steve Esomba, 2012). It is also described as an interface between classical engineering and biology (Ralf Neurohr and Cristian Dragomirescu, 2007).

Another definition of bionics has been given as, “Bionics is the scientific discipline, which is in charge with the systematic transfer of construction, process and evolution principles of living systems into technical applications”. (Neumann, 1993). A further definition of bionics was offered by Nachtigall who defined the field as, “Bionics is learning from nature for the creation of independent technical solutions”. (Nachtigall W, 2003).

Looking at a threefold approach, it explores the biological principles within technical systems, a second definition looks at the replacement of certain parts of the body with prostheses or implants: artificial mechanics yet another third process is exploring and interpreting the rules of nature using technical methods (U. Lindemann and J. Gramann, 2004).

A GLIMPSE OF HISTORY OF BIONICS

In the sixteenth century, Leonardo da Vinci sketched designs for gliding and flapping machines primarily based on his anatomical take a look at of birds.

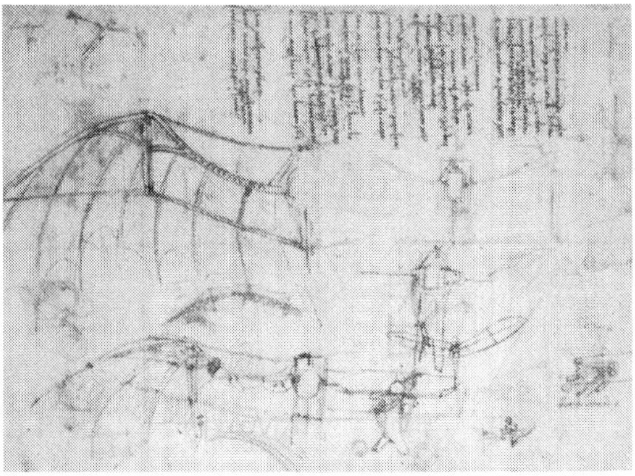

A Leonardo da Vinci sketch of a flying machine using a bird model

Source: http://library.acropolis.org/biomimicry-human-creation-inspired-by-nature/

After Da Vinci, more than three centuries later, Otto Lilienthal constructed and tested machines that could fly modelled on birds (Chanute, 1997). Another notable example of inspiration from nature is that of Orville and Wilbur Wright who constructed and manoeuvred aircrafts based on observation of buzzards (Wilbur and Orville, 1987).

Archaeological studies point to the use of artificial limbs into the annals of history. (Martijn Hendrix et al., 2016). An ancient prosthetic foot made of wood and leather was found in Egypt in 2000 that suggests its use.

An ancient prosthetic foot from Egypt

Source: http://www.ancientpages.com/2014/02/20/prehistoric-evidence-sophisticated-prosthetics-ancient-egypt-artificial-toes-helped-egyptians-walk/

In another example from history, Marcus Sergius, a general of the Roman emperor lost his hand in the second Punic war that led to the design of a prosthetic made of iron that even facilitated the operation of weapons and ultimately his participation in subsequent wars (Kulley, 2003).

The development of prosthetics was facilitated by The American Civil war (1861-1865) that led to several amputations. It has been reported that James Edward Hanger, a Confederate soldier lost a leg and was not satisfied with a peg that was a common replacement. This led to the foundation of Hanger Limb that could operate flexibly at the knee joint and the establishment of Hanger organization as a major player in this field.

Following the Second World War, the US Federal government established the American National Academy of Sciences that founded the Artificial Limb Program which aimed to develop prosthetics Apart from the war, incidents such as steam engine and rail road accidents added fuel to the industry to develop further research (Ott et al., 2002).

For instance, Carnes Artificial Limb Company based at Kansas used the tagline, "The Carnes arm puts you on the pay roll" while railroad workers were shown to use prosthetic limbs of railroad workers .

Apart from such employment and aesthetic aims, the use of prosthetics was augmented with the development of Becker hand in many types along with a cosmetic glove that covered the same to look "normal".

The use of bionic limbs has been applied in the field of public sports such as: Dutch runner and swimmer Marlou 'Blade Babe' van Rhijn and the famous Oscar Pistorius. The latter was dubbed as 'Blade Runner' as he congenitally lacked limbs that caused the use of bionic limbs for running. Though he achieved many accolades in Paralympics, his interest to participate in Olympics was spurned by the governing body whose decision was overruled by Court of Arbitration for Sport.

(Martijn Hendrix et al., 2016). Pistorius performs his feats with theuse of carbon-fibre artificial limbs and holds the double amputee world record in the 100, 200 and 400 metres events (Camporesi S, 2008).

The applications of bionics in the medical industry are plenty ranging from a cochlear implant to research on retinal bionics. The development of SONAR using the mechanisms of bats is an interesting example of bionics. The use of cochlear implants has reportedly enhanced the signals of the auditory nerves that are connected to the CNS. The system uses the mechanism of the signals in a functioning ear system that could aid not only people who are hard of hearing but also to aid in deaf children to hear and also in the recognition and process of speech.

Another arm of medical bionics includes addressing vision problems of patients with retinal issues: retinitis pigmentosa and macular degeneration. The development of a microchip that deals with has been discussed (Rajesh Mehrotra et al., 2011).

CASE STUDY OF MEDICAL BIONICS

Retinitis Pigmentosa (RP)

Retinitis pigmentosa (RP) is a leading and debilitating cause of both central and peripheral vision loss. It is reported as a genetically diverse disease commonly affecting both eyes in a progressive fashion with no efficient treatment.

A 2004 pilot study examined the efficacy of an artificial silicon retina (ASR) microchip for the treatment of retinitis pigmentosa that is associated with loss of vision. (Alan Y. Chow et al., 2004).

The authors engineered the ASR microchip as a 2-mm-diameter silicon-based device that containing 5000 microelectrode-tipped micro-photodiodes and powered by incident light. The ASR microchip was implanted in right eyes of 6 patients with retinitis pigmentosa while the left eyes served as controls.

Vincent Chow and Dr. Alan Chow, inventors of the ASR®device and co-founders of Optobionics Corp.

2mm ASR® device lying on a penny

Source: http://optobionics.com/asrdevice.shtml

The research team observed that in a span of upto a year and a half no adverse effects were observed i.e. implant rejection, infection, inflammation, erosion, neovascularization, retinal detachment, or migration were not reported. On the contrary, they reported enhanced perception of brightness, contrast, color, movement, shape, resolution, and visual field size (Alan Y. Chow et al., 2004).

Silicon-Neuron Junction

This involved the construction electronics and neuronal networks: silicon elements and nerve cells. They constructed a direct silicon-to-neuron junction: they stimulated an individual Retzius cell of the leech and reported that, "Extracellular stimulation of an individual neuron is possible from a microscopic silicon element with perfectly insulating silicon oxide" (Peter Fromherz and Alfred Stett, 1995).

WHY BIONICS?

A key reason for bionics to garner interest is the greater understanding of life: the knowledge as to the functioning of animals and plants to microbes. The concepts of cellular and sub-cellular molecules and metabolism have contributed to the understanding of the various complexities of life sciences. The sequencing of genomes has contributed to a huge thrust in the understanding of life mechanisms. Many mechanisms that were previously a mystery were unravelled that include the mechanism by which drag is reduced by penguins or the walking over water by certain lizards. These studies aid in the understanding of analogue issues. The use of bionics or understanding the mechanisms of the biological world can be extrapolated of the same to various methods in the world of mechanics and engineering.

Certain key noteworthy figures have explained various principles such as Galileo who applied the concepts of physics to elucidate the differences in the sized of the limbs of small and big mammals. Certain techniques such as atomic-force microscopy, particle–image velocimetry, x-ray cinematography, high-speed video have been attributed in the understanding of various principles. Such principles find application in the use of mechanics that uses the essential components of biological sciences.

Another factor that fuelled the research in the field of bionics includes advancements in materials and molecules that are used to fabricate biological tissues. The combinations of engineering, polymers, chemistry and molecular biology have ushered in the fabrication of various molecules used in the field of bioengineering. Various materials such as nanomaterials or micro- electromechanical devices that combine chips with biological sciences have aided in the development of the field of bionics (Dickinson, 1999).

BIONICS AND NATURE

Several examples of the use of biological structures and applications or products that are in use include: Velcro: George de Mestral invented this application when removing burrs off his dog's fur after walking in the Swiss countryside. The pattern of the burrs inspired him to design hooks and loops that were complementary to each other to fasten securely.

Burr from which Velcro was inspired

The trick in bringing Velcro to the masses was not only the working of a burr but also fabrication and mass-production of the fuzz and hook surfaces for its success (Velcro S.A, 1955 and Dickinson, 1999).

Sharks show the presence of ridges that are parallel to their body: this presence of grooves causes a decrease in the drag experienced by the animal, especially known swim fast. The use of the idea of shark skin aided the construction of riblet sheets that were used for the construction of fuselage and wings of an airbus. This resulted in the formation of Airbus 320 that has a lower usage of fuel.

The leaf of a lotus plant is associated with stagnant and often turbid or muddy water. Despite the lack of cleaning mechanisms, the leaves are not affected by water. This is attributed to minute protuberances on the surface of the leaf that are coated with wax. Unlike other leaves where water spreads on the surface of smooth leaves, these protuberances cause water to accumulate dust and fall off the leaf. This principle is used for Lotusan, ISPO paints that use this effect to clean a home from the effects of rains through the use of a "Lotus-Effekt" that has protected by Intellectual Property Rights (Dickinson, 1999).

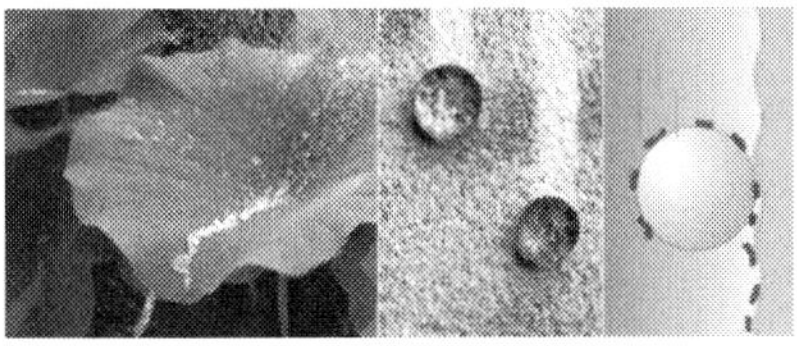

The principle of lotus leaf on the development of self-cleaning paint

http://www.farbenklick.de/15-fassadenfarbe-lotusan-lotusang-von-sto

The textures of various surfaces have been constructed to mimic such biological systems in order to repel dirt or even to facilitate its easy removal have been reported from several other leaves and insect wings (Wagner et al., 1996).

EXAMPLES OF NATURE INSPIRED BIONICS

A few more examples are presented that deal with nature inspired teams that propelled human inventions:

Geckos possess good climbing capacities that are based on submicrometre keratin hairs on their soles. Each of these hairs produce a force attributed to van der Waals and/or capillary interactions in an order of 10(-7)N that when combined together possess an adhesive force of 10

N x cm(-2). This force enables the geckos to remain attached to surfaces that include glass. Using this model, a research team in 2003 created a new type of adhesive by mimicking the gecko mechanism. They reported the successful generation of a 'gecko tape' made by microfabrication of dense arrays of flexible plastic pillars, whose geometry was optimized to ensure their collective adhesion (Geim A.K et al., 2003).

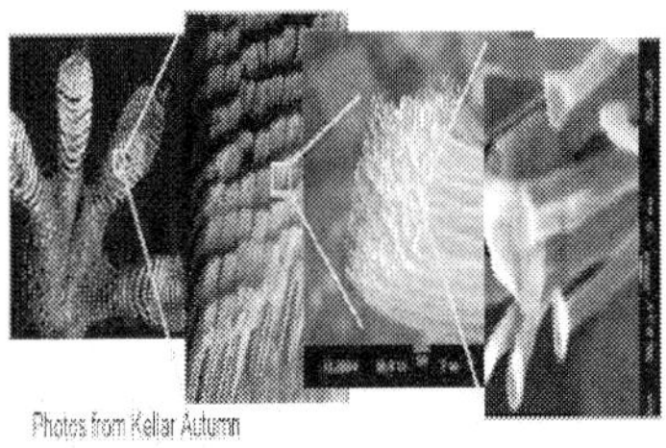

Comparison of the hair fibers of a gecko with that of the generated gecko tape

http://newatlas.com/gecko-inspired-adhesive-tape-market/41047/

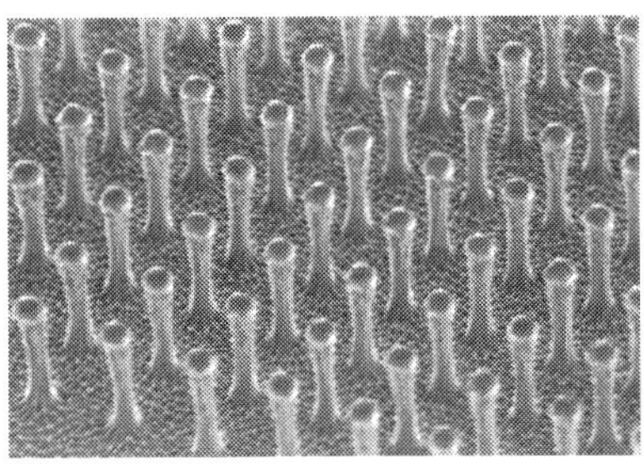

https://www.newscientist.com/article/dn3785-gecko-tape-will-stick-you-to-ceiling/

In another work, with a view of developing tough materials, the cause of tough wood in tension was attributed to the arrangement of cellulose along the cells: reported as at 15° to the long axis of the cells in softwoods. This arrangement was reproduced using glass fibers that were arranged in varying combinations in order to produce sturdy material (Gordon J.E, Jeronimidis G, 1980).

The same authors in a new team reported a patent where the structure developed by mimicking cellulose in wood as glass within resin could generate material that was really durable (Chaplin R.C, Gordon J.E, Jeronimidis G, 1983).

- Early history points to design of stable wings using the idea of Alsomitra macrocarpa that has winged seeds. These 15 cm wings aid in the flight of the seed over considerable length. (Coineau Y, Kresling B, 1987).
- The phenomenon of anti-reflection has been reported in several natural systems such as insect wings, eyes and the leaves of trees that inhabit the floor of tropical forests. Such a system has been replicated on a polythene sheet that is fixed to a glass on a solar panel system that has caused an increase of 10% in the absorption of light.
- Several creatures that inhabit aquatic systems have been explored in the field of bionics that include: penguins with very little drag exerted on their anatomy used in the design of dirigibles. Using the idea of propulsion by fish fins, such constructions have been called nektors used in submarines by Nekton, Inc.
- There are reports as to the trabecular struts in the head of the human femur, or the taper of a tulip stem that inspired the construction of the Eiffel tower. The monument was so constructed in order to resist wind and the curve of the pylons was so aligned in order to resist winds in accordance to the force and moment.
- Apart from the tower, the Crystal Palace was purportedly constructed using giant water lily leaves as a model that was displayed by the architect Joseph Paxton, at the Royal Society of the Arts in London. However, it is seen that there is a huge difference between radial tapering beams of a water lily and corrugations of the roof of the Palace (Vincent et al., 2006)
- Fast-swimming sharks have scales with microgrooves referred to as riblets aligned in the route of fluid glide. Riblets result in water shifting effectively over the surface. In preceding experiments and models, it had been shown that riblets provide drag discount through lifting vortices fashioned in turbulent float lowering average shear stresses. Riblets have proven drag reductions on the order of 10% whilst compared to a flat surface. To optimize riblet geometries for low drag, 3 exclusive geometries were modeled and their drag residences and vortex

structures were as compared. In addition, a shark-inspired geometry with riblets arranged in a scale pattern changed into modeled to examine shark scales to these riblet geometries. Through these models, the most beneficial riblet geometries and dimensions were decided (Martin and Bhushan, 2016).

- A research paper explored the potential of earthworm, a ubiquitous organism belonging to annelid. The animal possesses unique body features such as its non-smooth body surface or its pattern of locomotion. Special emphasis was placed on the combination of its surface lubrication, the rough surface and its charecterestic movement in the reduction of drag for bionic applications. (Gu Yunqing et al., 2016).

BIONICS AND INDUSTRIES

Using models of plants and animals several inventions have surfaced in the recent inventions of the industry. A few examples include:

- ***Bionic car***: Tropical boxfish have a movement resistance only 0.06 was used for the construction of a bionic car by Mercedes. This engineered bionic car reports of a resistance coefficient Cd = 0.19 as compared to Cd = 0.27 of conventional car from series production. Apart from the decrease in resistance encountered, the car also consumes lesser oil and possesses Selective Catalytic Reduction (SCR) to minimize the damage to the environment.

The Mercedes bionic car and boxfish

http://news.nationalgeographic.com/news/2005/06/0615_050615_fishcar.html

In another example, **Pylon Radar Cross Section** was engineered to be light weight on the basis of bamboo and bulrush. It was reported that the mass of the pylon was reduced by 43% while strength efficiency and specific stiffness efficiency was boosted by 52.9% and 43.6%. Thus, the use of bionics in the construction of lightweight materials has shown promising results.

Bionics was employed to decrease the mass of a 5-axe gantry type **milling machine** called Crossbeam structure of Lin MC6000 machining center. Based on the structures of the veins of leaves of giant water lilies, Gloeospermum leaf veins and Mexico cacti stem an improved design of ribs were obtained. Further, bionics helped in the reduction of weight by ~3.31%, along with decrease in deformation of 16.22%. The bionic model also aided in anti-vibrational properties by increasing the natural frequencies.

Several thin walled cylindrical structures find use in the industry such as civil, oil structures, aeronautical equipment and tanks in increasing order of their radii and thickness. Using a bionic model of a cross section of a bamboo stem, it was possible to show that "load bearing efficiency (kN/kg) for the bionic design is ~124.8% higher than for conventional one".

A Precise grinding machine-tool bed and column was used for the evaluation of a bionic model in order to improve its anti-vibrational properties along with enhanced stiffness and reducing the distortion. Based on the venation patterns of several leaves, bionic models were generated that offered better properties than the conventional machines. It was reported that along with an increase in stiffness of 12.3% , the deformation of the machine reduced by around 12% in the bionic model as compared to that of conventional construction (Adam Ruszaj, 2015).

BIONICS APPROACHES

Two approaches or strategies have been described for the field of bionics that include:

- **Topdown approach** that deals with looking at a defined goal or issue. This is used for scouring through analogues in nature that could provide a potential solution. The system of nature that is being examined as a solution is subjected to intense analysis to study the ways this process can be interpreted in engineering or mechanics. To quote, "The whole process of development and optimization can be seen as a dialog between the worlds of technology and biology. Though, there may be some early results, which appear very similar to the original proposal of nature, but the shape of the final technical solution (like the blended winglet) will in many cases not being traced back to its natural origin" (Ralf Neurohr and Cristian Dragomirescu, 2007).

An example used to illustrate this approach is the use of a blended winglet: the use of bionics to lessen loss of energy due to vortex at aeroplane wing tips. Based on the behaviour of wing tips of birds in flight, the technique can be applied to the development of an aeroplane wing with the desired aim.

In a nutshell, the steps involved in this approach have been reviewed:

1. Following proper definition of the aim, a suitable biological model or process is subjected to analysis and primary evaluation.
2. This is followed by a mathematical modelling (most ubiquitously used method is the Finite Elements Method) of the mechanical model of the biological entity.
3. Various parameters of the material to mimic the biological substance or process are evaluated such as stress, distortion, resistance to wear and fatigue etc (Adam Ruszaj, 2015).

- The other is the **bottoms up** strategy that begins at a novel discovery of life sciences that has been elucidated in detail. This principle is then applied to an "abstract, formal and interdisciplinary description of the phenomenon, which is opening the pathways for the transfer into different fields of technology".

  An example used to illustrate this mechanism includes the use of a drop of water on a lotus leaf that has led to the application of several self cleaning products and paints (Ralf Neurohr and Cristian Dragomirescu, 2007).

TOOL USED: CASE STUDY

Approximately 50 years ago in Russia, a in particular a success trouble-solving system commenced to be developed. It changed into named TRIZ, the acronym of Teorija Reshenija Izobretatel'skih Zadach (loosely translated as 'Theory of Inventive Problem Solving'). TRIZ is widely known for its success switch of various innovations and answers from one field of engineering to another. Since the principle thrust of science is likewise to transfer functions, mechanisms and ideas from one field to every other, TRIZ seems the correct start line.

A research team additionally used TRIZ as a useful précis and definition of engineering method, a novel use of the device. We realize of no different approach or system that's so powerful and so fashionable. Since TRIZ is not very widely recognized to Western technological know-how and technology, a quick description is necessary, outlining its everyday use through problem solvers.

TRIZ is a group of equipment and techniques, advanced by way of Genrich Altshuller and Rafik Shapiro (Altshuller 1999) that ensures accurate definition of a trouble at a useful degree and then affords strong indicators towards successful and often exceptionally innovative answers.

In many regions, the winter temperatures pass underneath 0°C, main to dangerously icy road conditions. But the spikes or chains regularly connected to wheels for the complete winter damage ice-unfastened roads. It is inconvenient to be changing always among unique iciness and summer time tyres or putting chains on and off the wheels. It might be better to have a right away converting tyre, which would be a conventional rubber tyre on an ice-unfastened street and capable of generate high friction on an icy floor.

Thus, they›ve a normal TRIZ conflict, with necessities to be soft and clean and to be stable, tough and sharp. It is likewise possible to formulate the problem as: 'they need ok friction between a wheel and a road below variable road floor conditions.' The friction need to also vary without the load of the car changing. Now follow the stairs above and try to find a biomimetic solution. This suggests that sharp and gentle components ought to be segmented and/or alternately structured (arranged in space), possibly as more than one claws and pads.

These alternating gadgets need to be alternated in time as nicely, in different phrases, gentle and sharp modes of operation of the wheel ought to be switched on and rancid in time. And in the end, IP 17 certainly indicates the need to design some spikes (claws, serrations, teeth and so on) to provide adequate grip for the wheel. All this shows that they should design the wheel/tyre with spikes, which will perform like cat›s claws—dynamically, in line with the prevailing street conditions.

The wheel and tyre already possesses IP 29 Pneumatics and hydraulics and IP 30 Flexible shells and skinny films—the flexible pneumatic rubber tyre. IP 2 Taking out have to be additionally employed as already advocated above. It means that the alternated segmented parts due to their dynamics should carry out the 'tender' and 'sharp' modes because of the removing precept. And the closing (but no longer least) point is that one has to pay attention to IP thirteen. The different manner round. It means that the meant wheel with the retractable/protractible claws is a great deal less difficult to layout vice versa—the rigidly set up claws are combined with the inflatable/deflatable soft component, the pneumatic tyre (Vincent et al., 2006).

USE OF STRUCTURAL DETAILS

The use of biological mechanisms such as nozzles by spiders to spin their silk contributes to the properties of spider silk. Drag line silk produced by the arthropods has been credited with several properties such as its strength compared with alloys or polymers and its ability to be stretched up to 30 percent. The potential of spider silk in parachutes or ropes or vests holds much promise. Apart from the cloning and expression of silk proteins in heterologous systems, it is important to examine the biological mechanisms used for the production of the silk by the animal that contributes to its slew of activities (Dickinson, 1999).

SUMMARY

The imperative precept of bionics may be summed up below the motto of 'studying from nature'. In the course of evolutionary process, nature has over tens of millions of years evolved several solutions that even nowadays confront scientists and engineers with all their advanced understanding. Nature's solution includes robust compounds, state-of-the-art mobility mechanisms, functional ways of evolution and living. As a result of the dynamic traits in organic studies and technology in particular at the molecular degree absolutely new approaches will emerge inside the discipline of bionics. Though the precise functions of biological aspects require further elucidation, the field of bionics nevertheless offers several potentials such as medical applications and industrial applications.

REFERENCES

1. Adam Ruszaj. (2015). Bionic Impact On Industrial Production Development. Advances In Manufacturing Science And Technology. 39(4), DOI: 10.2478/amst-2015-0019.
2. Alan Y. Chow, MD; Vincent Y. Chow, BS; Kirk H. Packo, MD; et al. (2010). The Artificial Silicon Retina Microchip for the Treatment of VisionLoss From Retinitis Pigmentosa. Arch Ophthalmol. 122(4):460-469. doi:10.1001/archopht.122.4.460.
3. Camporesi S. (2008). Oscar pistorius, enhancement and posthumans. J Med Ethics. 34(9), 639.
4. Chanute O. Progress in Flying Machines. New York: Dover; 1997.
5. Chaplin RC, Gordon JE, Jeronimidis G. Development of a novel fibrous composite material. 1983. USA Patent no. 4 409 274.
6. Coineau Y, Kresling B. Les inventions de la nature et la bionique. Museum National d'Histoire Naturelle; Paris, France: 1987.
7. Dickinson MH. Bionics: Biological insight into mechanical design. Proc Natl Acad Sci USA. 96(25):14208-14209.
8. Geim AK, Dubonos SV, Grigorieva IV, Novoselov KS, Zhukov AA, Shapoval SY. (2003). Microfabricated adhesive mimicking gecko foot-hair. Nat. Mater. 2, 461–463.
9. Gordon JE, Jeronimidis G. (1980). Composites with high work of fracture. Phil Trans R Soc A. 294, 545–550.
10. Howard F. (1999). Wilbur and Orville: A Biography of the Wright Brothers. New York: Knopf; 1987.
11. Kulley M. (2003). biomed.brown.edu. Retrieved from http://biomed.brown.edu/Courses/BI108/BI108_2003_Groups/Hand_Prosthetics/history.html.

12. Lee D.W. Unusual strategies of light absorption in rain-forest herbs. In: Givnish T.J, editor. On the economy of plant form and function. Cambridge University Press; Cambridge, UK: 1986. pp. 105–131.
13. Lindemann U, Gramann J. Engineering Design Using biological principles. International Design Conference - Design 2004. Dubrovnik, May 18 - 21, 2004.
14. Martijn Hendrikx, Gerko Schaap, Jeroen Wagenaar En Hans Mulder (2016). The Challenges Of Bionics. Volume I, Issue 1.
15. Martin S, Bhushan B. Modeling and optimization of shark-inspired riblet geometries for low drag applications.J Colloid Interface Sci. 2016 Jul 15;474:206-15. doi: 10.1016/j.jcis.2016.04.019. Epub 2016 Apr 16.
16. Nachtigall W., "Bionik", 2nd ed., Berlin – New York: Springer, 2003.
17. Neumann D., "Technologieanalyse Bionik. Analyse und Bewertunge zukünftiger Technologien.", VDI-Technologiezentrum, PhysikalischeTechnologien, Düsseldorf, Germany, 1993.
18. Ott K, Serlin D, Mihm S. (2002). Artificial Parts, Practical Lives: Modern Histories of Prosthetics, 45(4), 359. http://doi.org/10.1353/tech.2004.0187.
19. Peter Fromherz and Alfred Stett. (1995). Silicon-Neuron Junction: Capacitive Stimulation of an Individual Neuron on a Silicon Chip. Physical Review Letters. 75, 8, 1670-1673.
20. Rajesh Mehrotra, Sandhya Mehrotra and Jitendra Panwar. (2011). Bionics: Where machines imitate life. Short Communication. Biotechnol Bioinf Bioeng. 1(2):275-277.
21. Ralf Neurohr and Cristian Dragomirescu. Bionics in Engineering - Defining new Goals in Engineering Education at "Politehnica" University of Bucharest. International Conference on Engineering Education – ICEE 2007. Coimbra, Portugal.
22. Twenty-First Century's Fuel Sufficiency Roadmap By Dr. Steve Esomba, Published 2012
23. Velcro S.A. (1955). Improvements in or relating to a method and a device for producing a velvet type fabric. Patent no. 721 338 Switzerland.

24. Vincent JF., Bogatyreva OA, Bogatyrev NR, Bowyer A, Pahl A-K. (2006). Biomimetics: its practice and theory. J Royal Soc Interface. 3(9):471-482. doi:10.1098/rsif.2006.0127.
25. Vincent JF., Bogatyreva OA, Bogatyrev NR, Bowyer A, Pahl A-K. (2006). Biomimetics: its practice and theory. J Royal Soc Interface. 3(9):471-482. doi:10.1098/rsif.2006.0127.
26. Wagner T, Neinhuis C, Barthlott W. (1996). Wettability and contaminability of insect wings as a function of their surface sculptures. Acta Zool. (Stockholm). 76, 213–225.
27. Yunqing G, Tianxing F, Jiegang M, Lanfang J, Denghao W, Shuihua Z. (2016). A Review of Bionic Technology for Drag Reduction Based on Analysis of Abilities the Earthworm. International Journal of Engineering Research in Africa. 19, 103-111.

CHAPTER 2

BIONICS AND CASE STUDIES

Shiv Sanjeevi Sripathi[1] and Prerna Pandey[2]

[1]Shiv Sanjeevi Sripathi completed his BSc in Biotechnology from Mumbai University and secured second rank in Mumbai University and first rank in his college: Kishinchand chellaram College. Shiv secured first rank in his college KET's V.G.Vaze College and has worked on a stem cell project at the Specialized Centre for Cell Based Therapy (SCCT), KEM Hospital at Mumbai, following which he was a Junior Research Fellow at Microbiology & Cell Biology Department, Indian Institute of Science, Bangalore from August 2009 to November 2011. The field of work involved cloning of cell wall genes and transcription factors in E. coli & M. smegmatis. This was followed by a stint as a Research and Development Executive at Transasia Biomedicals, Mumbai where he worked on patent writing and developing immunoassays from 2011 till 2013. In 2013, he entered the teaching field where he has been tutoring biology and chemistry to the students of 11th and 12th grades along with English grammar and communication skills to people of all age groups.

[2]Dr Prerna Pandey, a biotechnologist with several years of wet lab research experience, worked at International Center for Genetic Engineering and Biotechnology, New Delhi. Her field of research in PhD included isolation and molecular characterization of geminiviruses, genome sequencing, gene annotation, and gene silencing using the RNA interference technology. She has also worked at Transasia Biomedicals and Advance Enzyme Technologies as a scientist. Prerna has published papers in peer-reviewed journals, and has submitted a number of annotated Geminiviral genome sequences in the GenBank, including two novel ones. She has also completed her editing and proof-

reading courses from Society for the promotion of Editors and Proof-readers, UK and now works as a freelance scientific editor and writer. When Prerna is not busy with her assignments, she loves to explore historical places.

INTRODUCTION

The development of gear to treat injury and disorder has been one in all mankind's earliest legacies. Luminaries inclusive of Leonardo da Vinci who ignited the development of engineering principles that might be implemented to the human frame (Artif Organs, 1999).

In 1958, health practitioner Jack Steele first proposed the time period bionics to account for the technique of looking at nature to solve the problems of engineering and mechanics (Vincent et al., 2006).

Advances in substances sciences, electrical engineering, biochemistry, stronger biomaterials, robotics, tissue engineering and computing power have led to large trends in medical bionics. As a end result, this area is now acknowledged as an vital issue of cutting-edge biotechnological healthcare treatments

Common examples of bionic treatment plans include mechanical limb prostheses, artificial muscles and hearts, retinal and cochlear implants. Though these are not replacements, they can provide many abilities. For instance, athlete Oscar Pistorius who has no legs, yet can outperform different the so called normal athletes, with the use of his carbon fibre leg prostheses (Camporesi S, 2008).

BIONICS FOR HEART

The effect of a bionic device called CorCap Cardiac Support Device (CSD; Acorn Cardiovascular, Inc, St. Paul, MN) especially its long term effects were tested by a research team in 2007. Such devices had been shown to reverse remodeling associated with such cardiac failures and support the structure and function of the left ventricle in such patients.

The team reported that such a device caused reduction of LV end diastolic volume and LV end systolic volume when compared to controls. It was also observed that implantation of CorCap concomitantly with mitral value surgery or by itself could improve the working of the ventricle in cases of heart failure (Starling et al., 2007).

BIONIC PANCREAS

Pancreas transplantation has been a successful remedy of sufferers with diabetic complications for years, constrained however to fairly small range of sufferers not only because of organ scarcity but additionally because of ischemic injury of retrieved organs. More than 50% of transplanted islets are lost throughout the primary few days put up transplant. It is possible that one of the causes of this is IBMIR - instant blood-mediated inflammatory reaction. Researchers controlled to 3D - bioprint vessels with diameter of 0.5 mm diameter, which were physiologically energized – vitamins and oxygen had been viable to be transported through the wall and cells of endothelium of that vessels produced steady junction among them. Improving of three-D bio printing era may help to construct scaffolds in order to be capable of being used in clinical medicine. Of the maximum significance seems to be engineering a Human Bionic Pancreas with islets and extracellular matrix that may help become islet transplantation a way of remedy for definitely larger group of patients with diabetes. Implementation of such Bionic pancreas could lower prices paid on the remedy of diabetes mellitus and its worry and can reverse terrible developments (Michal Wszola et al., 2015).

BIONICS AND EYE IMPLANTS

It was reported that Globally, the numbers of people with blindness is between 32 and 39 million people (Lewis PM et al., 2015). Among myriad techniques, to restore vision in the blind, a promising and common technique is bionic eye'. It works on the electric stimulation of the optic pathways: The underlying premise of such stimulation is to rouse neuronal pastime at a domain inside the visible pathway that remains functional regardless of the underlying motive of blindness. This propagates alongside the remaining intact visual pathway to the visible cortex, resulting in some form of visual notion. There are nearly 20 research groups developing retinal prostheses with varying implant places, with businesses based totally in Australia, USA, Germany, France and Japan.

Currently, there are many research centers that deal with visible prostheses, protecting each feasible target area inside the visual pathway. For instance, there is The Argus II device (Second Sight Medical

Products, Inc., Sylmar, CA, USA) and the Alpha IMS (Retina Implant AG, Reutlingen, Germany). These instruments have been approved for clinical use in the European Union in 2011 and 2013 respectively. The former device was approved by FDA in 2013 for Humanitarian Use Device program (Lewis PM, Ayton LN, Guymer, 2016).

An Overview of Prosthesis for the Eye

Visual bionic implants work on the recording of images by a camera that are attached on to glasses. The highlighting of objects, texts or floors is done by mechanisms such as feature extraction or Simplification of images. Following this, electric impulses are generated that are transmitted via wired or wireless connections.

Such an approach may also have a difference by the use of light-sensitive elements such as photodiodes that are either linked with the implant or combined with an array. The transmission of image data to photodiode/electrode array forms part of a strategy. In another mechanism, light on a photodiode array generates weak photoelectric signals that are amplified. This method offers advantages such as circumventing head movements to observe a place and facilitating fixation of natural gaze (Lewis PM, Ayton LN, Guymer, 2016).

Retinal implants have shown the largest numbers of human clinical trials up to now among the use of bionics in the field of vision. A study in 2014 conducted a trial using suprachoroidal implant in three subjects. This study by Bionic Vision,Australia reported that phosphenes (light percepts that are elicited in response to electrical stimulation) were more sustained apart from strong implants in vicinity (Ayton et al., 2014).

A 2015 study involved implantation of Argus II in patients with age-related macular degeneration (AMD) whose effects are expected in the future.

The field of bionic implants in the eye is being augmented by the use of several techniques to stimulate nerves apart from electrodes that include low-intensity centered ultrasound, alternating magnetic fields, using microfluidic devices to inject neurotransmitters, changing ionic gradients across neural cellular membranes and optogenetics.

There are even advancements in the connections between electrodes and the target neurons. A group based in Japan cultured neurons on a microelectromechanical substrate. These nerve cells interface anxious gadget neurons such as those in the retina without any delay. A noteworthy effect of the use of retinal stimulation includes delaying photoreceptor degeneration, anti-inflammatory and anti-apoptotic effects along with preservation of retinal ganglion cells following experimental damage.

Patients who were administered retinal prosthesis had reportedly bare light perception or minimum base line vision. The use of retinal bionics has caused an improvement in vision as studied by grating acuity tasks or Landolt C optotypes (Lewis PM, Ayton LN, Guymer, 2016).

EXOSKELETONS AND NEURODEGENERATIVE DISEASES

What Are Exoskeletons?

An exoskeleton is a distinctive kind of robot to be worn as an overall or frame, effectively supporting, or in some cases substituting for, the user's own movements (Yang et al., 2008). Exoskeletons are controlled by the user's movements and do not need any external control terminal (with the exception of a service terminal). The main parts are: the frame; the power system and the control system with sensors. Generally, an exoskeleton is a technical tool that expands and improves selected abilities of the user.

There are reports of exoskeletons with multiple control systems such as the HAL5(Hybrid Assistive Limb 5) exoskeleton with 2 controls: Cybernic Voluntary Control, based on bio-electric signals observed on the surface of the skin, and the Robotic Autonomous Control System, based on a database of elementary movements, allowing for exoskeleton movements despite poor bio-electric signals from the user.

Why Look at Exoskeletons?

From a medical point of view, as an alternative to wheelchairs, providing mobility and increasing a patient's physical abilities and also to support caretakers in tasks requiring major physical effort, e.g. and also gaining a better understanding of human body posture and movement.

Examples of exoskeletons in use include: WPAS (Wearable Power Assist Suit) by K. Yamamoto, Kanagawa Institute of Technology, Japan or RoboKnee, B.T.Krupp, C.J. Morse,Yobotics Corporation, USA for medical purposes. HULC (Human Universal Load Carrier), Lockheed Martin and Berkeley Bionics, USA and XOS Exoskeleton of Sarcos Research Corporation and Raytheon, USA are exoskeletons for military purposes (Emilia Mikołajewska, Dariusz Mikołajewski, 2011).

Potential of Exoskeletons

- These exoskeletons can enable a patient all-day supported mobility, with a full range of motions and proper movement patterns; the possibility of training in (or re-learning of) all functional activities:lying, sitting, standing, walking, transfer, stair climbing, ascending/descending slopes and activities of daily living.
- They can also provide efficient support for weak patients, improving their muscle strength, bone density, cardiovascular system and endurance, to prepare them for normal functioning; the possibility of nearly-natural all-day functioning, with posture and movements that are better for the metabolism and exercises.
- Generally, the use of exoskeletons can: increase the possibilities and effectiveness of rehabilitation, especially neurorehabilitation, through intensive all-day functional therapy during normal life activities; increase the accessibility of rehabilitation, including home rehabilitation and home care; increase the user's safety (fall protection); save money and time, decreasing the average hospital stay and the number of therapists needed per patient (Emilia Mikołajewska, Dariusz Mikołajewski, 2011).

How Do Exoskeletons Work?

Usually, the system of helping the consumer with a meant motion is as follows: evaluation of the current situation: posture, limb positions, etc., primarily based on indicators accumulated from EMG or other sensors (muscle contraction, and so on.) when the person attempts to move: the exoskeleton manipulate device analyzes the sensor indicators and determines the consumer's intended motion the manage system selects a pre-programmed motion sample, adjusts it to the person's cutting-edge position and supports the movement with actuators the use of an appropriate pressure four) after finishing the motion, the manipulate machine analyzes a brand-new situation such as posture, limb positions, assessing and preparing for the user's subsequent feasible moves (Fleischer et al., 2005, Suzuki et al., 2005 and Fleischer, 2007).

Additional applications attributed to such exoskeletons consist of balance manipulate and coordination (Cass, 2008).

What Are the Exoskeletons in Use?

Examples of exoskeletons in use encompass: WPAS (Wearable Power Assist Suit) by K. Yamamoto, Kanagawa Institute of Technology, Japan or RoboKnee, B.T.Krupp, C.J. Morse,Yobotics Corporation, USA for medical purposes. HULC (Human Universal Load Carrier), Lockheed Martin and Berkeley Bionics, USA and XOS Exoskeleton of Sarcos Research Corporation and Raytheon, USA are exoskeletons for military purposes (Emilia Mikołajewska, Dariusz Mikołajewski, 2011).

Clinical Uses of Exoskeletons

These exoskeletons can allow a patient all-day supported mobility, with a full variety of motions and proper movement styles; the possibility of training in (or re-learning of) all useful activities: mendacity, sitting, status, on foot, transfer, stair mountain climbing, ascending/descending slopes and activities of each day living.

They can also provide green help for susceptible sufferers, enhancing their muscle power, bone density, cardiovascular device and staying power, to prepare them for regular functioning; the possibility of almost-natural all-day functioning, with posture and movements which are higher for the metabolism and sporting events.

Generally, the use of exoskeletons can: enhance the possibilities and effectiveness of rehabilitation, in particular neurorehabilitation, via extensive all-day useful therapy at some point of regular existence sports; boom the accessibility of rehabilitation, consisting of home rehabilitation and home care; growth the user's protection (fall safety); store time and money, lowering the common sanatorium live and the quantity of therapists wanted in step with a patient (Emilia Mikołajewska, Dariusz Mikołajewski, 2011).

Case Study of Exoskeleton

The sudden inability to walk is a huge impediment following spinal cord injury (SCI) that reports an incidence of 10.4-83 per million (Wyndaele M, Wyndaele JJ, 2006). Apart from Traditional over ground (TOG) therapy, interest was gained in body weight supported training (BWST) to aid in the treatment. In order to evaluate the efficiency of robotic exoskeletons, a team in 2013 evaluated EksoTM. The team reported that the wearable EksoTM appeared to have potential in patients without volitional lower extremity function. It was reported to be safe for those with complete thoracic SCI, when utilized in a controlled environment with assistance by trained personnel. There were also no reports of pain while using the exoskeleton or other adverse effects such as no known fractures, swelling or skin effects. The use of these exoskeletons is suggested for those patients with defects in extremity function (Stephanie A Kolakowsky-Hayner et al., 2013).

AUTOBIONICS

The present-day loss of organic organs available for transplantation and the continued difficulties of immunosuppresion have caused our want for the extended improvement of bionic treatment options. Many of those 'traditional' bionic therapies in large part focus on digital or artificially engineered remedy answers. The use of these external artificial cures can therefore be taken into consideration as 'Exo-bionic' remedies (Greek exo - outside). Here human remedy and enhancement can take location via the re-corporation and re-layout of inherent organs and tissue within each character – the system of Auto-bionics.

Pancreas and Autobionics

Roux-en-Y Gastric Bypass (improving insulin sensitivity and release) – The burgeoning pandemic of global obesity coupled with terrible consequences of weight loss pharmacotherapy brought about the improvement of 'so-referred to as' weight reduction or bariatric techniques to deal with morbid weight problems. It was soon realized but that those strategies did a long way greater than without a doubt weight reduction. Pories and colleagues posted on their series of gastric bypass in obese patients after a 14 yr observe-up, reporting that in patients with type II diabetes mellitus, 83% carried out every day glucose degree within the absence of medication (Pories Wj et al., 1995). This became remarkable at the time, as the improvement in glucose metabolism befell within a few days of surgical operation before any noticeable weight loss.

The Roux-en-Y procedure creates a small belly pouch because of surgical partition of the higher a part of the stomach, with a rearrangement of alimentary organs that lets in processed meals to 'skip' the primary part of the belly whilst preserving the hormonal presence of the belly inside the alimentary tract and also preventing toxic biliary reflux.

These operations gain weight reduction by using decreasing appetite and meals intake however also alter metabolic rate with most effective minimal impact calorie absorption. Their dramatic effect on diabetes has been proposed to happen from two fundamental principles: hindgut hypothesis: One considers ingested vitamins bypassing the stomach which subsequently releases an altered 'gut hormonal signal' inside the decrease gut or foregut speculation: a lack of vitamins in the principal body of stomach may additionally bring about 'gut hormonal signals' launched from the upper gastrointestinal system.

The intestine hormones which can be notably improved by way of those operations include Glucagon-like peptide-1 (GLP-1) and Peptide YY (PYY).

GLP-1 might also have an critical function in the decision of diabetes:

- As it protects against pancreatic Beta-Cells Against Apoptosis (Cornu et al., 2009)
- Stimulates cAMP-established insulin exocytosis (Holz, 2004)
- It acts on other systems together with the heart in making contributions to the beneficial opposite remodelling seen inside the heart after those operations (Ashrafian et al., 2008).

In patients who expand submit-operative high insulin and low sugar, the application of this treatment resulted in a higher level of insulin to prevent many of these patients from seeking pancreatectomy to facilitate levels of the hormone (Ashrafian et al., 2010).

Cardiac Autobionics

Donald Ross finished the primary pulmonary valve autograft to replace an aortic valve in a human in 1967, a system that has seeing that been named after him (Ross, 1967).

The Ross Procedure (replacing the aortic valve) - Synthetic exo-bionic cardiac valve alternative is fraught with long term anticoagulation issues, especially in younger patients. In order to overcome these problems, experimental fashions brought using the pulmonary valve to update the aortic valve. Both valves had been morphologically semilunar with three leaflets, and had similar function that permitted them to be considered as substitutes for each other.

The treatment offers relief and facilitates cardiac haemodynamics and ventricular remodelling. A challenge in this treatment is the skill and technicality associated with the process The Ross system is administered to patients for treatment of aortic valve ailment, ladies who desire to bear children after this operation and also critical aortic stenosis in children (Ashrafian et al., 2010)

The use of Dynamic Cardiomyoplasty has been credited with several advantages. It entails harvest of left latissimus muscle and subsequently detached from the humerus. It is then wrapped across the cardiac ventricles and pulsed via electrodes linked to an electrical pacemaking tool. Both the geometry and the pacing device allow the muscle graft to agreement with the ventricles in synchrony to the coronary heart for the duration of cardiac systole. Several advantages reported include: acceleration of contraction of myocardial muscle, a decrease of stress on left ventricular wall and remodeling of opposite ventricles with beneficial effects.

To aid in the treatment of congestive cardiac failure, the process of aortomyoplaty can wrap skeletal muscle around the ascending or descending thoracic aorta. The method boasts of a process of less technical complications and early research has indicated that the process can aid in the assistance as obtained in an Intra-Aortic Balloon Pump (Sherwood et al., 2002).

Autobionics and Anal Muscle Repair

Gracilis is a muscle of the thigh with minimum contribution to locomotion in humans that propelled studies to test its use in several surgeries that involve reconstruction (Macchi et al., 2008).

In dynamic graciloplasty, the gracilis muscle is explanted and then transposed to encircle the anal canal. The aim of this manner is to bring about modifications inside the rapid-twitch fatigable (Type II) muscle fibres of gracilis to convert them into sluggish-twitch, fatigue-resistant (Type I) muscle fibres that extra carefully resemble the histological characteristics of the inner anal sphincter. The gracilis in its regular country only has 43% kind I fibres rather than the anal sphincter which has 80% kind I fibres.

The muscle is 'wrapped' across the anal canal forming either alpha, gamma (most commonplace) or epsilon loops to reflect the affected person's pelvic anatomy. An electric neurostimulator located inside the tissue of the lower stomach wall, and communicates with the gracilis muscle thru electrodes. The neurostimulator discharges a modern-day that contracts the muscle (for the majority of the day), but whilst stimulation is turned off, the muscle relaxes to facilitate defecation.

The use of autobionics along with implanting a synthetic bowel sphincter (ABS) report similar values of continence though the latter method is suggestive of 82 % to 100 % improvement in comparison to 74-82.8 % of the graciloplasty (Belyaev et al., 2006).

It is suggested that the use of bionic sphincters along with gene therapy or stem cell therapy can aid in the therapy of such disorders (Ashrafian et al., 2010).

Skeletal Muscle Bionics

This process involves improvement of the action of Skeletal Muscle Ventricles where an accessory cardiac pump is formed by using skeletal muscle from latissimus dorsi ventricles. The latissimus dorsi muscle is mobilized and wrapped as a conical spiral around a mandrel to be is left inside the chest hollow space. This is followed by electrical conditioning to enable the wrap to become resistant to fatigue. The device can now serve as an accessory cardiac pumping chamber. Certain advantages reported on this technique include that the incidence of rupture is decreased

(Thomas et al., 1996) and also do not cause additional complications like anti-clotting (Thomas et al., 1996).

Other models reported include a pump that produces power equivalent to 22 % of left ventricular electricity and is implanted in the left atrium and aorta (Hooper et al., 1992).

Research also indicates the advantages of such bionic devices in animal models. It was reported that they also allow seeding of autogenous endothelial cells. Their output is reported as about half of that of the left ventricle after a period of 4 hours. Their contribution to helping the heart is similar to mounted intra-aortic balloon pump (IABP) (Ashrafian et al., 2010).

SUMMARY

The use of bionics in several medical fields such as eye, heart, and pancreas were discussed. The development of exoskeletons: seen in several elements of nature in the assistance of debilitating disorders were evaluated.

The field of bionics in retinal implants has been promising so have the use of the novel field of autobionics. The use of several techniques in the latter approach such as Graciloplasty, Cardiomyoplasty, Ross procedure and Skeletal Muscle Ventricles are the tip of the iceberg. With several diseases plaguing the world the use of latest technology such as bionics offers promising results to help meet the goal of biotechnology: use of science for betterment of mankind.

REFERENCES

1. Ashrafian H, Darzi A, Athanasiou T (2010) Auto-bionics – a new paradigm in Regenerative Medicine and Surgery. Regenerative Medicine 5(2):279-288. doi:10.2217/rme.10.2.
2. Ashrafian H, Le Roux Cw, Darzi A, Athanasiou T (2008) Effects of bariatric surgery on cardiovascular function. Circulation 118(20):2091–2102.
3. Ayton LN, Blamey PJ, Guymer RH et al (2014) First in human trial of a novel suprachoroidal retinal prosthesis. PLoS One 2014; 9:

e115239.

4. Belyaev O, Muller C, Uhl W (2006) Neosphincter surgery for fecal incontinence: A critical and unbiased review of the relevant literature. Surg Today 36(4):295–303.
5. Camporesi S (2008) Oscar pistorius, enhancement and post-humans. J Med Ethics 34(9):639.
6. Cass AB: Preliminary specifications for an exoskeleton for the training of balance in balance impaired individuals (2008) MSc Degree Thesis. Virginia Polytechnic Institute and State University.
7. Cornu M, Yang Jy, Jaccard E, Poussin C, Widmann C, Thorens B (2009) Glp-1 protects beta-cells against apoptosis by increasing the activtiy of an igf-2/igf1-receptor autocrine loop. Diabetes 58(8):1816-1825.
8. Fleischer C (2007) Controlling exoskeletons with EMG signals and a biomechanical body model. PhD Thesis. Technische Universität Berlin.
9. Fleischer C, Reinicke C, Hummel G (2005) Predicting the intended motion with EMG signals for an exoskeleton orthosis controller. Proc 2005 IEEE Int Conf Robot Auton Syst (IROS), 2029–2034.
10. GG Holz (2004) Epac: A new camp-binding protein in support of glucagon-like peptide-1 receptor-mediated signal transduction in the pancreatic beta-cell. Diabetes 53(1):5–13.
11. Hooper Tl, Niinami H, Hammond Rl, et al (1992) Skeletal muscle ventricles as left atrial-aortic pumps: Short-term studies. Ann Thorac Surg 54(2):316–322.
12. Kolakowsky-Hayner S, Crew J, Moran S, Shah A (2013) Safety and Feasibility of using the EksoTM Bionic Exoskeleton to Aid Ambulation after Spinal Cord Injury. J Spine S4. http://dx.doi.org/10.4172/2165-7939.S4-003.
13. Leonardo da vinci M (1999) Precursor member of the international society for rotary blood pumps? Artif Organs 23(6):477–479
14. Lewis PM, Ackland HM, Lowery AJ, Rosenfeld JV (2015) Restoration of vision in blind individuals using bionic devices: a review with a focus on cortical visual prostheses. Brain Res 1595C: 51–73.

15. Lewis PM, Ayton LN, Guymer RH, et al (2016) Advances in implantable bionic devices for blindness: a review. Anz J Surg 86(9):654-659. doi:10.1111/ans.13616.
16. Macchi V, Vigato E, Porzionato A, et al (2008) The gracilis muscle and its use in clinical reconstruction: An anatomical, embryological, and radiological study. Clin Anat 21(7):696–704.
17. Mikołajewska E, Mikołajewski D (2011) Exoskeletons in Neurological Diseases-Current and Potential Future Applications. Adv Clin Exp Med 20(2): 227–233.
18. Pories Wj, Swanson Ms, Macdonald Kg, et al (1995) Who would have thought it? An operation proves to be the most effective therapy for adult-onset diabetes mellitus. Ann Surg 222(3):339–350. discussion 350-332.
19. Ross DN (1967) Replacement of aortic and mitral valves with a pulmonary autograft. Lancet 2(7523):956–958.
20. Sherwood Jt, Schomisch Sj, Thompson Dr, George Dt, Cmolik Bl (2003) Aortomyoplasty: Hemodynamics and comparison to the intraaortic balloon pump. J Surg Res 110(2):315–321.
21. Starling Rc, Jessup M, Oh Jk, et al (2007) Sustained benefits of the corcap cardiac support device on left ventricular remodeling: Three year follow-up results from the acorn clinical trial. Ann Thorac Surg 84(4):1236–1242.
22. Suzuki K, Kawamura Y, Hayashi T et al (2005) Intention based walking support for paraplegia patient. Proc IEEE Int Conf on Systems, Man and Cybernetics 2005, 2707–2713.
23. Thomas Ga, Isoda S, Hammond Rl, et al (1996) Pericardium-lined skeletal muscle ventricles: Up to two years' in-circulation experience. Ann Thorac Surg 62(6):1698–1706. discussion 1706-1697.
24. Thomas Ga, Lu H, Isoda S, et al (1996) Skeletal muscle ventricles in circulation: Decreased incidence of rupture. Ann Thorac Surg 61(1):430–436.
25. Vincent Jf, Bogatyreva Oa, Bogatyrev Nr, Bowyer A, Pahl Ak (2006) Biomimetics: Its practice and theory. J R Soc Interface 3(9):471–482.

26. Wszola M, Idaszek J, Berman A, Kosik A, Gorski L, Jozwik A, et al (2015) Bionic Pancreas and Bionic Organs – how far we are from the success. MED tube Science III (3): 25 -27.
27. Wyndaele M, Wyndaele JJ (2006) Incidence, prevalence and epidemiology of spinal cord injury: what learns a worldwide literature survey? Spinal Cord 44:523-529.
28. Yang C-Y, Zhang J-F, Chen Y et al (2008) A review of exoskeleton-type systems and their key technologies. Proceedings of the Institution of Mechanical Engineers, Part C: J Mech Eng Sci 222(8), 1599–1612.

CHAPTER 3

DESIGN AND TESTING OF A BIONIC DANCING PROSTHESIS

Elliott J. Rouse[1], Nathan C. Villagaray-Carski[1,2], Robert W. Emerson[3], and Hugh M. Herr[1]

[1]Biomechatronics Group, MIT Media Lab, Massachusetts Institute of Technology, Cambridge, Massachusetts, United States of America

[2]Department of Electrical Engineering and Computer Science, Massachusetts Institute of Technology, Cambridge, Massachusetts, United States of America

[3]A Step Ahead Prosthetics, Burlington, Massachusetts, United States of America

ABSTRACT

Traditionally, prosthetic leg research has focused on improving mobility for activities of daily living. Artistic expression such as dance, however, is not a common research topic and consequently prosthetic technology for dance has been severely limited for the disabled. This work focuses on investigating the ankle joint kinetics and kinematics during a Latin-American dance to provide unique motor options for disabled individuals beyond those of daily living. The objective of this study was to develop a control system for a bionic ankle prosthesis that outperforms conventional prostheses when dancing the rumba. The biomechanics of the ankle joint of a non-amputee, professional dancer were acquired for the development of the bionic control system. Subsequently, a professional dancer who received a traumatic transtibial amputation in April 2013 tested the bionic dance prosthesis and a conventional, passive prosthesis for comparison. The ability to provide similar torque-angle behavior of the biological ankle was assessed to quantify the biological realism of the prostheses. The bionic dancing prosthesis overlapped with 37 ± 6% of the non-amputee ankle torque and ankle angle data, compared to 26 ± 2% for the conventional, passive prosthesis, a statistically greater overlap ($p = 0.01$). This study lays the foundation for quantifying unique, expressive activity modes currently unavailable to individuals with disabilities. Future work will focus on an expansion of the methods and types of dance investigated in this work.

INTRODUCTION

For decades, prosthetic design and control research has focused on enhancement of activities of daily living, such as ascending and descending stairs and inclines [1,2]. Such work has provided considerable advancement in community mobility and substantially impacts the quality of lives of these individuals. However, because prosthetic devices are designed for such common activities, more unique and expressive activities, such as dance, are cumbersome and unintuitive. Furthermore, little is known regarding the appropriate mechanical behavior required by the prosthesis during such activities. Recent improvements in bionic prosthetic devices have provided the opportunity to vary the mechanical characteristics of prosthetic joints in real-time [3,4]. These advancements have laid the foundation for prostheses that can accommodate a wide range of activity modes, including expressive activities, but the underlying control systems have yet to be studied.

The purpose of this study was to develop a control system for a bionic ankle-foot prosthesis capable of dancing the rumba, a popular Latin-American rhythm dance. Ankle joint kinematics and kinetics were obtained from a non-amputee professional dancer during various steps of the rumba. The data were used to develop a biologically inspired control system for the bionic ankle-foot prosthesis. Following testing, the ankle joint kinetics and kinematics of the bionic dance prosthesis were compared to those from a common passive prosthetic ankle worn while performing the rumba. The intent of this work was to extend the mobility of individuals with below-knee amputations to permit expressive activities that have traditionally remained difficult and unintuitive, reducing the gap in motor abilities between persons with and without leg amputation.

MATERIALS AND METHODS

Bionic Ankle Prosthesis

A commercially available bionic ankle prosthesis (BiOM Inc, Medford, MA) was modified for use in the study. The bionic device was chosen for its ability to provide joint position, impedance and torque control to the wearer, as well as the ability to implement novel control programs on the computers within the device. Two physical modifications were made to the bionic ankle-foot prosthesis. The prosthesis was modified to permit a lightweight spherical attachment between the prosthesis and the prosthetic socket. This enabled the prosthetist to easily adjust the alignment of the prosthesis. Additionally the spherical mount modification permitted a space within the socket where the battery was relocated. Relocating the battery provided a more anthropomorphic shape for the prosthesis and socket, allowing the experimental participant with leg amputation to perform cross-step dance maneuvers without leg-to-leg interference. The modifications of the prosthesis reduced the total mass by 100 g (Fig 1).

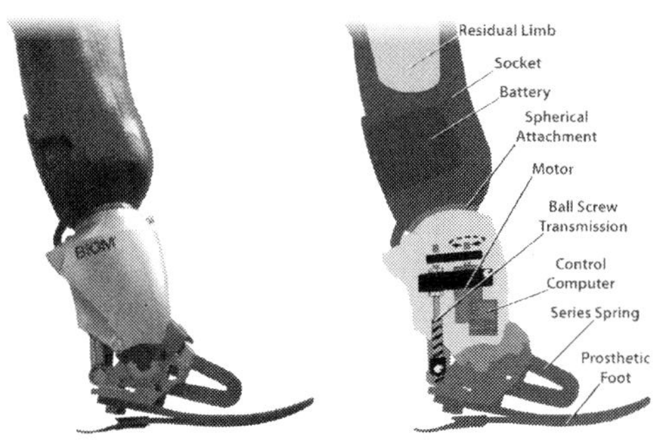

Figure 1. Schematic of bionic dancing prosthesis. Bionic ankle prosthesis shown (left) with major components highlighted (right). Note the location of the battery in the distal prosthetic socket.

Rumba Biomechanics of a Non-Amputee

Biomechanical data were obtained from a 39 year old, female professional dancer (height: 1.75 m, mass: 54.5 kg) in the gait lab within the Biomechatronics group at the MIT Media Lab. The study was approved by the MIT Committee On the Use of Humans as Experimental Subjects and written informed consent was obtained. Five trials were obtained of four steps of the rumba, with each trial lasting approximately 5–15 seconds. The specific steps were chosen because they span a broad range of the rumba dance. The four steps included in the study are known as basic, crossovers double-rock, crossovers with underarm turn, and open break. The participant was instructed to attempt to keep one foot on the force platforms within the gait lab, if possible, to maximize the data usable for the analysis.

During the dance steps, kinetic data were collected using the force plates within an instrumented treadmill (Bertec Corporation, Columbus OH) and two in-ground force plates (AMTI, Watertown, MA). Force data were sampled at 1 kHz while the kinematic data were collected synchronously via an infrared camera system (Vicon Motion Systems, Oxford, UK) and sampled at 100 Hz. The motion capture system utilized passive markers placed at 42 (bilateral) locations on the participant's body according to the Helen Hayes marker model. The raw data were processed in SIMM (Musculographics Inc., Evanston, IL) to obtain joint trajectory and torque profiles. Data were segmented to only include regions where one foot was acting on a single force platform. The relationship between ankle torque and ankle angle was explored for the development of the controller, where negative angles denoted plantar flexion.

Development of a Biologically-accurate Control System

The goal of the control system was to provide a biologically-accurate relationship between ankle torque and ankle angle during the rumba. The ankle torque for each step was investigated as a function of ankle angle while the angular velocity was greater than 30 degrees/s, chosen to eliminate regions of the dance steps where the ankle behavior was static. Linear regression was used to determine the mean relationship between torque and angle for each dance step. Subsequently, the slope and intercept were averaged across dance steps to obtain the mean linear torque-angle relationship that described the rumba dance steps. A linear spring-damper model of ankle mechanical characteristics was chosen for the desired behavior implemented in the control system, a model that has previously been shown to successfully characterize other human ankle behaviors [5,6]. The slope of the mean linear regression value corresponds to the stiffness and the equilibrium angle was determined by the x-intercept of the regression. In other words, the control system was defined by

$$\tau = k(\theta - \theta_0) + b\dot{\theta} \quad (1)$$

where τ corresponds to ankle torque, k is ankle stiffness, θ is ankle angle, θ_0 is equilibrium angle, b is the damping coefficient and $\dot{\theta}$ is ankle angular velocity measured by the bionic ankle. The damping value was tuned to the preference of the amputee wearer. Following development, the control system was downloaded to the control computer within the bionic dance prosthesis (Fig 1).

Testing with an Amputee Dancer

To quantify the performance of the biologically-inspired control system, a participant with a unilateral transtibial amputee tested the device. The participant was a 33 year old female (height: 1.72 m, weight: 54.5 kg) who lost her left leg in a traumatic amputation in April 2013. The participant is a trained, professional dancer. At the time of testing, she was 15 months post amputation and her residual limb was 40% of the length of her unaffected side (measured from the lateral epicondyle to the lateral malleolus). For daily use, she used a pin suspension with a patient adjustable Elation prosthetic foot (Össur, Reykjavik, Iceland). The participant has given written informed consent to publish these case details.

Prior to testing, the participant's prosthetist fit her with the bionic dance leg and aligned the device. She was permitted ample time to familiarize herself with the bionic device, and the experiment proceeded only when she reported being comfortable. The control system was modified based on feedback from the participant. During testing, she completed each of the four steps of the rumba in the gait lab. Five trials of each dance step were recorded using the aforementioned experimental protocol. Following testing with the bionic dance prosthesis, she repeated the trials with her conventional prosthesis. Identical post-processing and analysis were performed, as explained above.

To quantify the biological realism of the control system, the fraction of overlapping torque-angle areas was used. For each dance step, the combined torque-angle data from all trials were segmented into areas of 3° and 0.06 Nm/kg, making a grid of torque-angle regions. The number of regions that contained data from both the non-amputee dancer and the amputee participant was divided by the total number of regions in the non-amputee data. In other words, this fraction quantified the amount of overlap between the amputee participant's data and the non-amputee data. The fractions were calculated for both prosthesis test conditions (bionic dance leg and conventional passive prosthesis). The overlap percentages were compared with a paired, two-tailed t-test and the level of significance was set to 0.05.

RESULTS

Development of a Biologically-accurate Control System

The way ankle torque varied with ankle angle was used as the framework for the biologically-inspired control system. The results for the linear regressions of each step are shown in Table 1and a representative plot of the non-amputee torque-angle data is shown in Fig 2. Following feedback from the amputee participant regarding performance and comfort during the dance steps, the final values for the stiffness, equilibrium angle and damping were 0.011 Nm/°/kg, -10° and 9.7 x 10^{-4} Nms/°/kg, respectively.

Table 1. Regression results from non-amputee data.

Step	Slope (Nm/°/kg)	Intercept (Nm/kg)	θ_0	R^2
Basic	0.016	0.52	-32°	0.16
Double Rock	0.017	0.65	-38°	0.30
Under Arm	0.021	0.58	-27°	0.32
Open Break	0.018	0.63	-35°	0.27
Mean	0.018	0.595	-33°	

doi:10.1371/journal.pone.0135148.t001

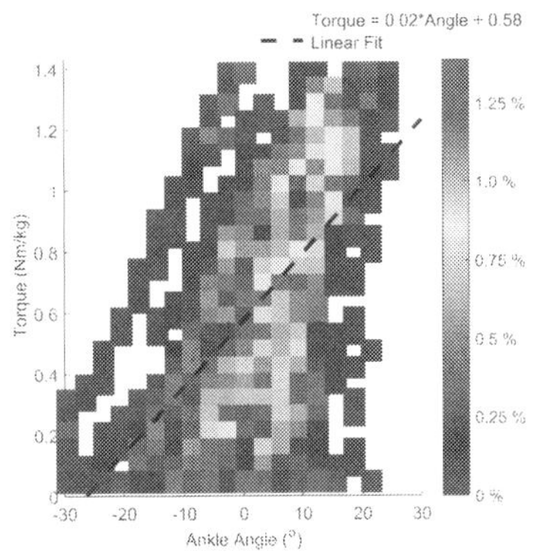

Figure 2. Torque vs. ankle angle data. Non-amputee ankle torque plotted against ankle angle for the under arm dance step. The color of the region indicates the percentage of the dance step data within the region; the darkest red regions had the highest concentration of data. The line of best fit is shown, which was used as the basis for the biologically inspired control system.

Comparison between Bionic and Conventional Prosthesis

The bionic control system was more biologically realistic than the passive conventional prosthesis. The passive conventional prosthesis overlapped with 26 ± 2% (mean ± standard deviation) of the non-amputee data, when averaged across the four dance steps. The bionic control system overlapped with 37 ± 6% of the non-amputee data regions (Fig 3). When compared across dance steps, the overlap of the bionic control system was statistically greater than the passive prosthesis ($p = 0.01$).

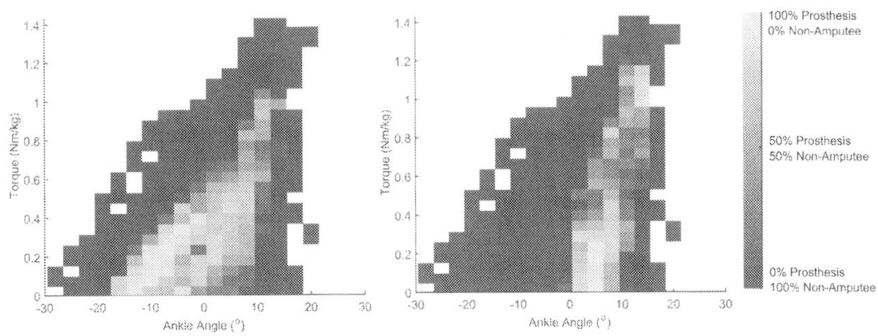

Figure 3. Overlap for prosthesis conditions. Ankle torque and angle data shown for the Crossovers dance step, with the bionic prosthesis condition shown on the left and the passive conventional prosthesis shown on the right. Color denotes the fraction of prosthesis data in the region compared to the non-amputee data.

DISCUSSION

The bionic dance prosthesis was designed to provide the appropriate relationship between ankle angle and ankle torque, during the steps of the rumba. It should be noted that the control system was not acting in a feed-forward manner, indicating that the amputee participant was not "riding" the bionic prosthesis. Instead, she biomechanically interacted with the prosthesis through her prosthetic socket, and the biologically-appropriate kinetic and kinematic relationship was enforced.

The bionic dance prosthesis significantly outperformed the conventional prosthesis when comparing kinetics and kinematics during the rumba; however, there is much work to be done. The range of motion for current bionic dance prostheses is designed based on walking kinematics. Dancing requires a larger range of motion than walking. This motivates the need for new bionic prostheses designed specifically for such uncommon motor activities such as dance.

The control system built for the bionic device assumes a single spring-damper model for the human ankle joint. As a result, no net-positive mechanical power is added by the bionic prosthesis during the dance steps. Future work will focus on a control system that provides appropriate levels of mechanical power to ankle joint, in a comfortable, biologically appropriate way. Additionally, to be able to replicate the area shown in Fig 2 (rather than a line), the equilibrium angle would need to be varied dynamically. The stationary nature of the control system demonstrated in this work is not able to provide time-varying dynamics. Future work may focus on more specialized control systems provide time-varying behavior and dance step recognition, improving the impact of the bionic device.

This study lays the foundation for quantifying other dances and expressive activity modes, rather than only activities of daily living. By understanding such activities, devices can be created that extend the motor skills available to amputees, improving their quality of life and diminishing the gap between the abilities of amputees and non-amputees.

AUTHOR CONTRIBUTIONS

Conceived and designed the experiments: EJR NCVC RWE HMH. Performed the experiments: EJR NCVC RWE. Analyzed the data: EJR

NCVC. Contributed reagents/materials/analysis tools: EJR NCVC RWE HMH. Wrote the paper: EJR NCVC RWE HMH.

REFERENCES

1. Hargrove LJ, Simon AM, Young AJ, Lipschutz RD, Finucane SB, et al. (2013) Robotic Leg Control with EMG Decoding in an Amputee with Nerve Transfers. New England Journal of Medicine 369: 1237–1242. pmid:24066744
2. Herr HM, Grabowski AM (2012) Bionic ankle–foot prosthesis normalizes walking gait for persons with leg amputation. Proceedings of the Royal Society B: Biological Sciences 279: 457–464. pmid:21752817
3. Au S, Herr H (2008) On the design of a powered ankle-foot prosthesis. The Importance of series and parallel motor elasticity. IEEE Robotics and Automation Magazine 15: 52–59.
4. Sup F, Varol HA, Mitchell J, Withrow TJ, Goldfarb M (2009) Preliminary evaluations of a self-contained anthropomorphic transfemoral prosthesis. Mechatronics, IEEE/ASME Transactions on 14: 667–676.
5. Kearney RE, Hunter IW (1990) System identification of human joint dynamics. Critical Reviews in Biomedical Engineering 18: 55–87. pmid:2204515
6. Rouse EJ, Hargrove LJ, Perreault EJ, Kuiken TA (2014) Estimation of Human Ankle Impedance During the Stance Phase of Walking. Neural Systems and Rehabilitation Engineering, IEEE Transactions on 22: 870–878.

CHAPTER 4

AN OVERVIEW OF BIONICS IN HUMAN MEDICINE AND ADVANCES IN GENERATION OF BIONIC HUMAN

Prerna Pandey[1] and Shiv Sanjeevi Sripathi[2]

[1]Dr Prerna Pandey, a biotechnologist with several years of wet lab research experience, worked at International Center for Genetic Engineering and Biotechnology, New Delhi. Her field of research in PhD included isolation and molecular characterization of geminiviruses, genome sequencing, gene annotation, and gene silencing using the RNA interference technology. She has also worked at Transasia Biomedicals and Advance Enzyme Technologies as a scientist. Prerna has published papers in peer-reviewed journals, and has submitted a number of annotated Geminiviral genome sequences in the GenBank, including two novel ones. She has also completed her editing and proof-reading courses from Society for the promotion of Editors and Proof-readers, UK and now works as a freelance scientific editor and writer. When Prerna is not busy with her assignments, she loves to explore historical places.

[2]Shiv Sanjeevi Sripathi completed his BSc in Biotechnology from Mumbai University and secured second rank in Mumbai University and first rank in his college: Kishinchand chellaram College. Shiv secured first rank in his college KET's V.G.Vaze College and has worked on a stem cell project at the Specialized Centre for Cell Based Therapy (SCCT), KEM Hospital at Mumbai, following which he was a Junior Research Fellow at Microbiology & Cell Biology Department, Indian Institute of Science, Bangalore from August 2009 to November 2011. The field of work involved cloning of cell wall genes and transcription factors in E. coli & M. smegmatis. This was followed by a stint as a Research and Development Executive at Transasia Biomedicals, Mumbai where he worked on patent writing and developing immunoassays from 2011 till 2013. In 2013,

he entered the teaching field where he has been tutoring biology and chemistry to the students of 11th and 12th grades along with English grammar and communication skills to people of all age groups.

INTRODUCTION

Human body is analogous to machinery dependent on the proper functioning of its various independent parts. Since the dawn of consciousness, malfunctioning of organs or their damage due to injuries and diseases had made people hoped for regeneration of their damaged organs or replacement with new ones. With the advancements in transplantation technologies, successful transplantation of several organs is observed. However, the field is plagued with various lacunae, the major one being the unavailability of recipient compatible organs.

Visions of the future where a man severely injured in a plane crash gets hope by artificial limbs and organs are now translated into reality. An intricate meshwork of biology, electronics and material sciences has given rise to the field of bionic medicine where prosthetic limbs and artificial organs are designed to mimic the functions of their counterparts present in human body. Integration of electronic devices with cells help in monitor the function, repair of damaged areas and even enhances the existing function of specific tissues and organs. Several new nanomaterials have been developed to provide better mechanics and making 3D printable organs an option. Moreover, in this era of bioengineering, we are closer more than ever to create a fully functional bionic human. This vision has been partially fulfilled with the creation of $1 million worth Rex aka Frank, which has created using organs developed by 17 different manufacturers. This will be discussed in brief in section 4 of the chapter.

Here, in this chapter we give brief insights into the technological advancements made in the field of bionic biomedicine. Moreover, we also give a brief overview of the organs and limbs developed. Recent developments in the field of 3D printing to print 3D organs are also discussed. We also discuss the closest creation to bionic man which is functional, Rex aka Frank in this chapter. This chapter aims to give a glimpse in the field of bionics applied in health and medicine.

BIONIC ORGANS: SENSORY AND NON-SENSORY

Both sensory (eye, ears, tongue) and non-sensory organs (heart, kidney, spleen, pancreas and limbs) have been designed artificially using the knowledge of electronics and human physiology. A few of the prototypes have been widely used, while others are in clinical trials. We briefly discuss some of both types of organs below.

BIONIC OPHTHAMOLOGY

Bionics has been applied in opthamology from creating an artificial retina to making glasses which allow blind people to see basic shapes, light and movement. Again, efforts in electronics, material sciences and surgery work hand in hand to achieve the desired effects.

Retinal pigmentosa and dry age-related macular degeneration is associated with photoreceptor-specific degeneration. In the past, gene therapy, stem cell related approaches and various drugs have been investigated. Certain cells in retina remain functional and can be stimulated to create the vision. Retinal based visual prostheses has taken two broad approaches namely epiretinal approach (Caspi et al., 2009; Eckmiller, 1997; Humayun et al., 2003; Rizzo et al., 2003; Roessler et al., 2009) and the subretinal approach, (Chow and Chow, 1997; Chow et al., 2004; Chow AY, May 21, 1991; Chow and Peachey, 1998; Peyman et al., 1998; Zrenner et al., 1999) based on the position of implanted electronic device relative to retina. In Epiretinal approach, a semiconductor device which powers a patterned electrode grid was placed in close contact with surface of nerve fiber layer of the retina (Caspi et al., 2009; Humayun et al., 1996; Humayun et al., 2003; Roessler et al., 2009). This system was used temporarily in six patients but was reported to cause mild inflammation and gliosis in certain patients (Roessler et al., 2009). Sterile hypopyan and epiretinal changes were observed in one of the patients respectively.

Subretinal approaches were initiated by Dr. Adam Chow and Mr. Vincent Chow in early 1990's (Chow and Chow, 1997; Chow et al., 2004; Chow AY, May 21, 1991; Peyman et al., 1998) . They had developed the Artificial Silicon Retina. This self-contained retinal implant was 2 mm in diameter, 0.3 mm thick and had thousands of microphotodiode cells across the area. This had a relatively simple design, where these microphotodiode cells generate electrical impulses which indeed activate the remaining functional retinal cells. Both hyperpolarized and depolarized currents are produced which enables the implanted patient to visualise light and darkness respectively. Several advantages associated with subretinal approach make it a preferred choice over epiretinal approach. Here, light acts as a power source for the device and thus, no external power sources like batteries are required. Furthermore, a relatively simple and low current device is used in this approach. Even surgeries to implant subretinal devices are simpler as compared to epiretinal devices.

A pilot clinical study was conducted with six Retinal pigmentosa patients. The ASR chip was implanted in the right eye of these patients and three patients were followed up for 6 months, while the remaining three were followed up for 18 months. Widely perceived areas of phosphenes were created by low voltages and currents of photodiodes induced by LEDs themselves. Another ASR chip implant design tested was composed of approximately 5000 P-N junction microphotodiodes pixels in N-i-P configuration. No adverse safety related effects were observed in six patients during 18 months of observation. All patients observed vision improvements of the implanted eye (Chow et al., 2004).

Recently, optical prostheses have been developed by Argus II Retinal Prosthesis and Harvard Research Fellow Dr. John Pezaris. The basic mechanism of the device is depicted in Figure 4.1.

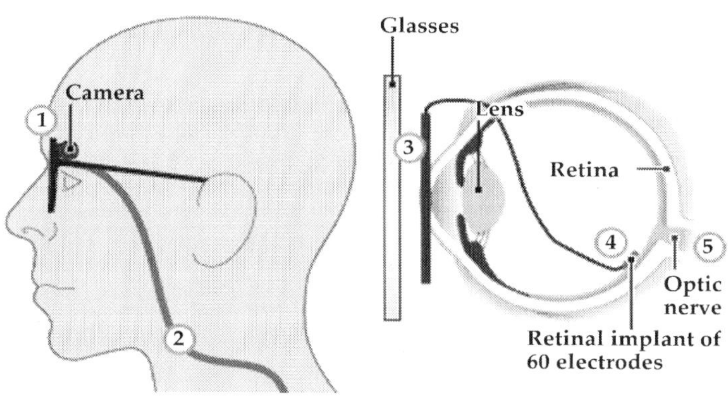

Figure 4.1. Design of the bionic eye. The eye consists of glasses with camera which captures the image, a processor which processes it and sends the signals to glasses and a receiver under the eye. Receiver sends information to elec-

trodes in retinal implant, which thereby stimulate the retinal cells to relay the information to the brain. Picture courtesy: http://news.bbc.co.uk/2/hi/science/nature/6368089.stm

The device developed by Argus II involves implantation of electrode in the eye, while that of John Pezaris is developed to send visual data straight to the brain bypassing the eye. Argus II is designed to create visual perception in blind individuals suffering from severe to profound retinitis pigmentosa (Stronks and Dagnelie, 2014).

COCHLEAR IMPLANTS

Djourna and Eryies had initiated the cochlear implants had been initiates in 1950s (Eisen, 2003). Placement of a coil wire in the ears of two deaf people had led to stimulation of auditory signals, however, the device had to be removed a while later. Modern cochlear implants employ different auditory signal route than normal and thus patients take time to adjust. In this period known as 'accommodation', 'acclimatization' or 'adaptation' spanning from 3-6 months, patients hears most of the words in unfamiliar sentences (Spahr A. J., 2004).

For patients with dysfunctional auditory nerve like those suffering from neurofibromatosis type 2 (NF2) and a Schwann cell disorder implants which cause direct stimulation of the brain were developed. This brain stem auditory implant can directly stimulate cochlear nucleus which is present in the brain stem. These are excited by the surface mounted 'button' electrodes. Varied specialized neurons in cochlear nucleus perform specific processing and feature extraction on relayed signals. Moreover, various tonotopic dimensions present in the nucleus can aid in activation of more than one pitch. Thus, the patients are able to detect and distinguish basic sound patterns like loudness, pitch and temporal resolution.

Auditory midbrain implants (AMI), which consists of an array of 16 electrodes placed on the surface (Mossop, 2000) or 20 electrodes that penetrate the inferior colliculus (Lenarz et al., 2006) have been developed recently. Pilot study conducted on six patients, resulted in support in lip reading, wide acceptance and normal psychophysical performance. Additionally, hybrid cochlear implants, which combine the properties of hearing aids and cochlear implants have been developed (Figure 4.2). They have been approved by the US Food and Drug Administration (FDA) for patients with severe sensorineural hearing loss of high-frequency sounds. This hearing loss may be caused due to old age, exposure to loud volumes, drugs or certain diseases and involves damage to the inner ear. The system developed, known as The Nucleus Hybrid L24 Cochlear Implant system depends on an external microphone and a speech processor, which collect sounds and convert them into electrical impulses. Hearing aid portion of the device amplifies the low frequency range sounds. These signals pass through small bundle of implanted electrodes and impulses are then transmitted to the cochlea. A clinical study with 50 participants revealed better word and sentence recognition after 6 months of the implant than conventional hearing aid. However, adverse side effects like tinnitus, electrode malfunction, low-frequency hearing loss and dizziness were reported in majority of the patients.

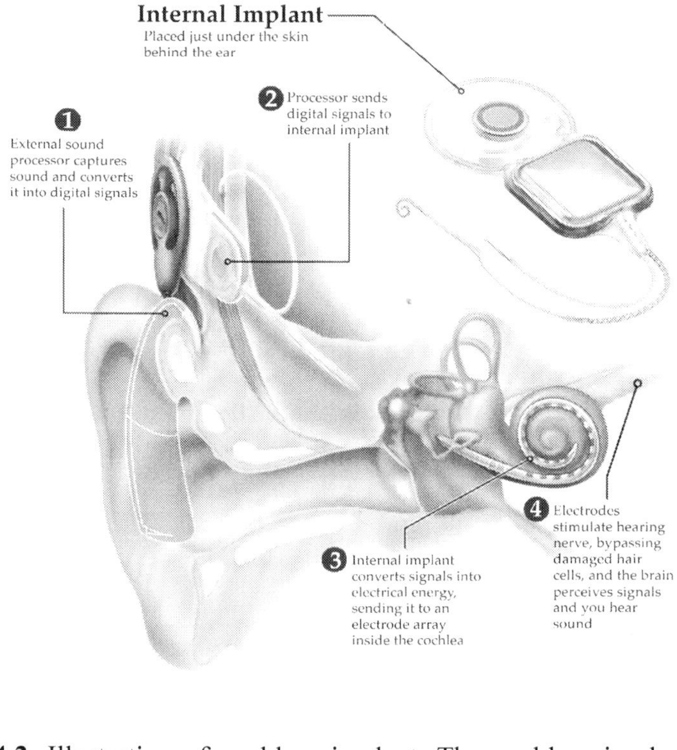

Figure 4.2. Illustration of cochlear implant. The cochlear implant depicted consists of external sound processor, which collects sounds and converts them into digital signals. These are relayed to internal implant by processor. Internal

implant converts digital signal to electrical signals which excite the electrode array in cochlea and hence resulting in stimulating the auditory nerves. Picture courtesy: http://www.kidshearnow.org/

PANCREAS

Most case of type I diabetes is categorized as an autoimmune reaction, where in body's own immune cells destroy the insulin-producing (islet) cells in the pancreas. Complications in these patients can be avoided by maintaining blood glucose levels under normal threshold. Inability to maintain these levels cause an increase in dose of treatment which increases the risk of life threatening acute and chronic complications in these patients. Need of the hour is a device which could constantly monitor the glucose and inject insulin accordingly.

Bionic pancreas had been developed for the management of type I diabetes (Russell et al., 2014). The system consists of bihormonal bionic system in which a thin needle used for monitoring glucose levels in tissue fluids is inserted under skin. A small wireless monitor along with pumps to administer insulin and glucagon is carried by patients (Figure 4.3). Pilot study using this device revealed modest success with significantly reduced blood glucose levels and slightly less reports of hypoglycaemia. Further improvements in this system are under development (Tauschmann and Hovorka, 2014).

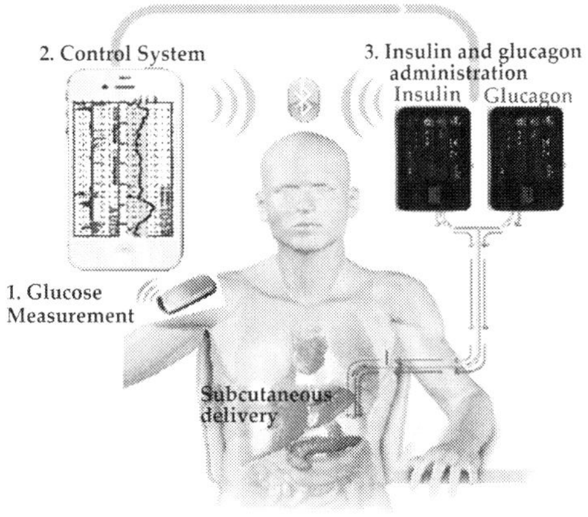

Figure 4.3. Representation of 'bionic pancreas' model. The representation depicts the bihormonal bionic pancreatic system with monitoring system which

measures the glucose levels in tissue fluids, a control system to relay the information to releasing pumps, where the two hormones insulin and glucagon are released subcutaneously in the body. Pic courtesy: http://fortune.com/2014/06/16/apple-powered-bionic-pancreas-one-step-closer/

A primary clinical trial (NCT02092220) was conducted using this device. 39 type I diabetes patients participated, of which 20 were assigned bionic pancreas, while 19 were assigned with either conventional therapy or sensor-augmented insulin pump therapy. Bionic pancreas was initially adjusted with body mass, and then later autonomously adaptive dosing algorithms were used to decide the amount of insulin delivered subcutaneously. Bionic pancreas proved to be more effective in glycaemic regulation without the requirement for carbohydrate counting (El-Khatib et al., 2017).

HEART

Numerous efforts are being invested in understanding the cellular and molecular mechanisms to understand the underlying mechanisms of cardiovascular diseases. These have aided in designing drugs, however, electronic devices to substitute improper functioning components of CV system have been developed parallel and have proved to be successful over the years.

Intraventricular conduction delay due to inefficient contraction and uncoordinated wall function has been observed in a subset of cardiac patients. The most popular treatment for this condition is implantable pacemaker. Pacemaker consists of batteries, computerized generator and wires with electrodes on the tips. Electrodes monitor the heart beat and send the data to the computerized generator. In case of abnormal heart rhythm, electric pulses are generated simultaneously to left and right ventricle. These have improved the symptoms and decreased the mortality (94). However, 30% of the patients who receive this treatment are not benefitted clinically, which may be attributed to different mechanisms of arrhythmia. It may be plausible that electrical stimulation delay and mechanical discoordination may both be contributing to the condition.

Implantable cardioverter defibrillator (ICD) was earlier conceptualized and developed for treating patients with ischaemic heart disease and who had suffered from ventricular arrhythmias. However, they were proved to be more useful in decreasing the mortality rate of at-risk patients of cardiac arrhythmia. Improvements in identification of at-risk patients using genomic, molecular and electrophysiological methods are underway.

Another device developed is the left ventricular assist device or LVAD, which has the potential to aid in functioning of left ventricle. This therapy has indicated an improvement in Ca^{2+} cycling, molecular and morphogenic remodelling and adrenergic responsiveness as reflected by the myocardium of the patients after implantation. Furthermore, a clinical trial wherein combined β2-AR-agonist treatment and LVAD therapy suggests that a total myocardial recovery is possible.

Non-excitatory cardiac contractility modulation involves generation of electrical stimuli to change heart function without any contraction. High frequency (10-Hz) electrical pulse is applied to heart during the electrical refractory period, which then improves heart function. The aim of this therapy is to improve native cardiac contractility for long term effects.

KIDNEY

Dialysis has been the solution for many kidney related problems for many years, but has various limitations like difficulty in implementation and interference in routines of patients. These may be slightly overcome by the development of wearable hemodialysis machines also known as wearable artificial kidney or WAK. These may provide advantage of miniaturization, mobility, and patient-oriented management. These systems will integrate dialysis membrane, vascular access, patient monitoring system, power sources and pumping systems for blood and fluids. Characteristics madatorily required in these systems would be safe connection/ disconnection of system and noninvasive monitoring to prevent biofilm and inflammatory effects.

Certain developments have been made in this field and one of the latest additions to WAKs is A peritoneal-based automated wearable artificial kidney (AWAK), which does not require extracorporeal circulation of blood (Lee and Roberts, 2008). This prevents the steps of additions of anticoagulation and continuous monitoring to avoid complications associated with blood-artificial membrane interaction. In this, a sorbent based system is used to regenerate aqueous (AqC) and the protein (PrC) constituents of the spent dialysate. This purified AqC has slightly high pH then physiological level, contains bicarbonate but has the same composition as the available peritoneal dialysate. Other sorbents can be added to remove additional toxins. The prototype under construction is hoped to have much higher dialysate flow rate (approximately 8-12 times the current ones).

Another system developed is Vicenza wearable artificial kidney for peritoneal dialysis (ViWAK PD) (Ronco and Fecondini, 2007). This system is designed to provide continuous ambulatory peritoneal dialysis (CAPD) for chronic kidney disease patients. Essentially, the device consists of a dialysate regeneration circuit, a double lumen peritoneal catheter, filter for deaeration, dialysate outflow and inflow lines and a computerised remote control. It was observed that clearance of creatinine and beta2-MG was in the range of 15-16 liters/day. Thus, this offers a possible alternative to CAPD and APD (automated peritoneal dialysis) as it decreases the number of exchanges and consumption of fluids when compare to respective techniques. Again, improvements in terms of glucose and bicarbonate injection system, reduced delivery of fibrin to sorbent and standardization of sorbent mixture to ensure complete removal of small molecules are required.

A step closer to create a continuous renal replacement therapy specially designed for neonates and small infants is the development of Cardio-Renal Pediatric Dialysis Emergency Machine (CARPEDIEM). This technology provides feasibility of treatment to neonates with acute kidney damage where peritoneal dialysis is not possible. CARPEDIEM was used on a 2.9 kg neonate suffering from haemorrhagic shock, multiple organ dysfunction, and severe fluid overload. Improvements in symptoms of fluid overload and severe acidosis were observed. Moreover, creatinine and bilirubin concentrations were also managed. Organ function restoration was reported; however mild renal insufficiency was still present but did not require renal replacement therapy (Ronco et al., 2014).

3D PRINTING OF ORGANS

Conventional electronic devices are two-dimensional, which prevents proper integration of functional components into biological tissue and organs. Approaches which can enable this integration can play instrumental role in regenerative medicine and prosthetics. 3D printing aims at creating three dimensional organs or body parts. During the computer aided design models, hypothetical layers are made and these are then built upwards using human cells.

Recent efforts in this directed by Mannoor et al. had generated fully functional bionic ears. They had used a cell seeded hydrogel matrix in the presence of conducting polymers with infused silver nanoparticles to generate 3D bionic ear (Mannoor et al., 2013). The ear consisted of cochlea shaped electrodes, inductive coil antenna, which allows the read out of inductively coupled signals from electrodes and the cultured cartilage tissue. The authors were able to generate both left and right functional ears that exhibited enhanced auditory sensing.

Other achievements include the development of two-dimensional tissue like skin and cartilage (https://3dprint.com/162193/wfirm-3d-printed-skin-burn-care/, (Cui et al., 2012), hollow tubes like vasculature (Marga et al., 2012), aortic valve (Duan et al., 2013) and tracheal splint (Zopf et al., 2013) and an early stage kidney prototype (http://www.wakehealth.edu/Research/WFIRM/Our-Story/Inside-the-Lab/Bioprinting.htm).

BIONIC HUMAN

Amalgamation of various individual organs (by 17 manufacturers) by roboticists Rich Walker and Matthew Godden of Shadow Robot Co.led to the creation of first fully functional bionic man named Rex or Frank. The feat was achieved in approximately three months' time and costed approximately $1 million. Functional organs in Rex include retinal and cochlear implant, a functioning heart, lungs, kidney, spleen and pancreas. He was physically made to resemble Bertolt Meyer, a social psychologist form Switzerland. The arms used are manufactured by i-LIMB made by Touch Bionics, while the robotic ankles and feet are from BiOM in Bedford, Mass. The Sources of various organs are mentioned in Figure 4.4.

SUMMARY

Much advancement is made in the field of bionics and these have made serious contributions in medical health care. Prosthetic limbs-arms, hands, legs and feet were the first to be made in this field. Recently, mind-controlled bionic arm has been developed and restored the touch. Apart from limbs, many organs as described earlier have been developed and paved their way to clinical trials. Sensory organs like bionic eye, cochlear implants and bionic tongue are developed which improve the functioning of patients who have suffered injuries/diseases and have lost their sensory capacities. Partial functioning capacity is restored to patients and it is hoped that further improvements in the field may even lead to full recovery of the organs and senses. Other organs like heart, kidney, pancreas, spleen, penis are also manufactured. These artificial organs are expected to revolutionize the modern medicine by making the organs available in customised format and also will be independent of the risk of immune rejection often observed in transplantation studies.

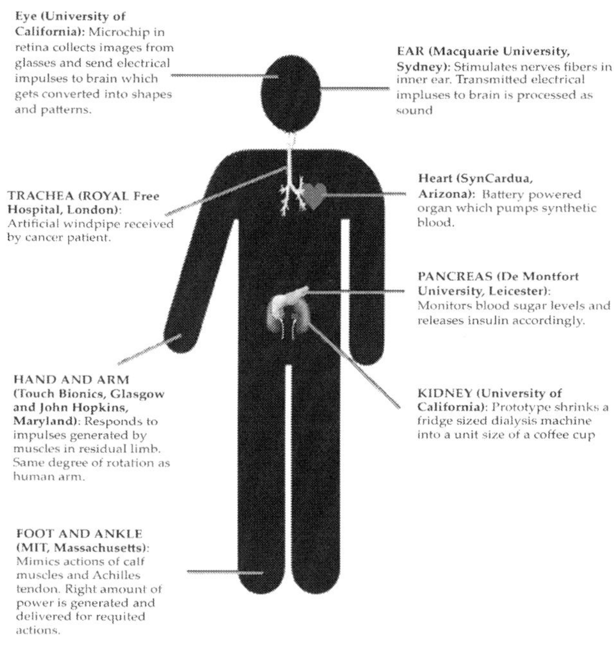

Figure 4.4. Illustration of Rex, the bionic human. The illustration describes a few functional organs present in the humanoid along with their sources and a brief description of each.

Important strides in hybrid organs are also been made in recent times, providing more opportunities for the development of compatible functional organs. 3D printing of organs that may prove an important arena for generation of tissues and organs needs to be explored further. All these developments in microengineering, sensor technology, material engineering and a greater understanding of human physiology has made a science fiction of a bionic man a feasibility today.

REFERENCES

1. Caspi, A., Dorn, J.D., McClure, K.H., Humayun, M.S., Greenberg, R.J., and McMahon, M.J. (2009). Feasibility study of a retinal prosthesis: spatial vision with a 16-electrode implant. Archives of ophthalmology *127*, 398-401.
2. Chow, A.Y., and Chow, V.Y. (1997). Subretinal electrical stimulation of the rabbit retina. Neuroscience letters *225*, 13-16.
3. Chow, A.Y., Chow, V.Y., Packo, K.H., Pollack, J.S., Peyman, G.A., and Schuchard, R. (2004). The artificial silicon retina microchip for the treatment of vision loss from retinitis pigmentosa. Archives of ophthalmology *122*, 460-469.
4. Chow AY, i. (May 21, 1991). Artificial retina device. US patent *5,016,633.*
5. Chow, A.Y., and Peachey, N.S. (1998). The subretinal microphotodiode array retinal prosthesis. Ophthalmic research *30*, 195-198.
6. Cui, X., Breitenkamp, K., Finn, M.G., Lotz, M., and D'Lima, D.D. (2012). Direct human cartilage repair using three-dimensional bioprinting technology. Tissue engineering Part A *18*, 1304-1312.
7. Duan, B., Hockaday, L.A., Kang, K.H., and Butcher, J.T. (2013). 3D bioprinting of heterogeneous aortic valve conduits with alginate/gelatin hydrogels. Journal of biomedical materials research Part A *101*, 1255-1264.
8. Eckmiller, R. (1997). Learning retina implants with epiretinal contacts. Ophthalmic research *29*, 281-289.
9. Eisen, M.D. (2003). Djourno, Eyries, and the first implanted electrical neural stimulator to restore hearing. Otology & neurotology:

official publication of the American Otological Society, American Neurotology Society [and] European Academy of Otology and Neurotology *24*, 500-506.

10. El-Khatib, F.H., Balliro, C., Hillard, M.A., Magyar, K.L., Ekhlaspour, L., Sinha, M., Mondesir, D., Esmaeili, A., Hartigan, C., Thompson, M.J., *et al.* (2017). Home use of a bihormonal bionic pancreas versus insulin pump therapy in adults with type 1 diabetes: a multicentre randomised crossover trial. Lancet *389*, 369-380.
11. Humayun, M.S., de Juan, E., Jr., Dagnelie, G., Greenberg, R.J., Propst, R.H., and Phillips, D.H. (1996). Visual perception elicited by electrical stimulation of retina in blind humans. Archives of ophthalmology *114*, 40-46.
12. Humayun, M.S., Weiland, J.D., Fujii, G.Y., Greenberg, R., Williamson, R., Little, J., Mech, B., Cimmarusti, V., Van Boemel, G., Dagnelie, G., *et al.* (2003). Visual perception in a blind subject with a chronic microelectronic retinal prosthesis. Vision research *43*, 2573-2581.
13. Lee, D.B., and Roberts, M. (2008). A peritoneal-based automated wearable artificial kidney. Clinical and experimental nephrology *12*, 171-180.
14. Lenarz, T., Lim, H.H., Reuter, G., Patrick, J.F., and Lenarz, M. (2006). The auditory midbrain implant: a new auditory prosthesis for neural deafness-concept and device description. Otology & neurotology : official publication of the American Otological Society, American Neurotology Society [and] European Academy of Otology and Neurotology *27*, 838-843.
15. Mannoor, M.S., Jiang, Z., James, T., Kong, Y.L., Malatesta, K.A., Soboyejo, W.O., Verma, N., Gracias, D.H., and McAlpine, M.C. (2013). 3D printed bionic ears. Nano letters *13*, 2634-2639.
16. Marga, F., Jakab, K., Khatiwala, C., Shepherd, B., Dorfman, S., Hubbard, B., Colbert, S., and Gabor, F. (2012). Toward engineering functional organ modules by additive manufacturing. Biofabrication *4*, 022001.
17. Mossop, J.E., Wilson, M.J., Caspary, D.M. & Moore, D.R. (2000). Down-regulation of inhibition following unilateral deafening. Hear Res *147*, 183–187

18. Peyman, G., Chow, A.Y., Liang, C., Chow, V.Y., Perlman, J.I., and Peachey, N.S. (1998). Subretinal semiconductor microphotodiode array. Ophthalmic surgery and lasers *29*, 234-241.
19. Rizzo, J.F., 3rd, Wyatt, J., Loewenstein, J., Kelly, S., and Shire, D. (2003). Perceptual efficacy of electrical stimulation of human retina with a microelectrode array during short-term surgical trials. Investigative ophthalmology & visual science *44*, 5362-5369.
20. Roessler, G., Laube, T., Brockmann, C., Kirschkamp, T., Mazinani, B., Goertz, M., Koch, C., Krisch, I., Sellhaus, B., Trieu, H.K., *et al.* (2009). Implantation and explantation of a wireless epiretinal retina implant device: observations during the EPIRET3 prospective clinical trial. Investigative ophthalmology & visual science *50*, 3003-3008.
21. Ronco, C., and Fecondini, L. (2007). The Vicenza wearable artificial kidney for peritoneal dialysis (ViWAK PD). Blood purification *25*, 383-388.
22. Ronco, C., Garzotto, F., Brendolan, A., Zanella, M., Bellettato, M., Vedovato, S., Chiarenza, F., Ricci, Z., and Goldstein, S.L. (2014). Continuous renal replacement therapy in neonates and small infants: development and first-in-human use of a miniaturised machine (CARPEDIEM). Lancet *383*, 1807-1813.
23. Russell, S.J., El-Khatib, F.H., Sinha, M., Magyar, K.L., McKeon, K., Goergen, L.G., Balliro, C., Hillard, M.A., Nathan, D.M., and Damiano, E.R. (2014). Outpatient glycemic control with a bionic pancreas in type 1 diabetes. The New England journal of medicine *371*, 313-325.
24. Spahr A. J., D.M.F. (2004). Performance of Subjects Fit With the Advanced Bionics CII and Nucleus 3G Cochlear Implant Devices. Arch Otolaryngol Head Neck Surg *130*, 624-628.
25. Stronks, H.C., and Dagnelie, G. (2014). The functional performance of the Argus II retinal prosthesis. Expert review of medical devices *11*, 23-30.
26. Tauschmann, M., and Hovorka, R. (2014). Insulin pump therapy in youth with type 1 diabetes: toward closed-loop systems. Expert opinion on drug delivery *11*, 943-955.
27. Zopf, D.A., Hollister, S.J., Nelson, M.E., Ohye, R.G., and Green,

G.E. (2013). Bioresorbable airway splint created with a three-dimensional printer. The New England journal of medicine *368*, 2043-2045.

28. Zrenner, E., Stett, A., Weiss, S., Aramant, R.B., Guenther, E., Kohler, K., Miliczek, K.D., Seiler, M.J., and Haemmerle, H. (1999). Can subretinal microphotodiodes successfully replace degenerated photoreceptors? Vision research *39*, 2555-2567.

CHAPTER 5

FIRST-IN-HUMAN TRIAL OF A NOVEL SUPRACHOROIDAL RETINAL PROSTHESIS

Lauren N. Ayton[1], Peter J. Blamey[2,3], Robyn H. Guymer[1], Chi D. Luu[1], David A. X. Nayagam[2,4], Nicholas C. Sinclair[2], Mohit N. Shivdasani[2,3], Jonathan Yeoh[1], Mark F. McCombe[1], Robert J. Briggs[1], Nicholas L. Opie[1], Joel Villalobos[2], Peter N. Dimitrov[1], Mary Varsamidis[1], Matthew A. Petoe[2], Chris D. McCarthy[5,6], Janine G. Walker[5,6], Nick Barnes[5,6], Anthony N. Burkitt[2,7], Chris E. Williams[2], Robert K. Shepherd[2,3], and Penelope J. Allen[1]

[1]Centre for Eye Research Australia, University of Melbourne, Royal Victorian Eye and Ear Hospital, East Melbourne, Australia

[2]Bionics Institute, East Melbourne, Australia

[3]Department of Medical Bionics, University of Melbourne, East Melbourne, Australia

[4]Department of Pathology, University of Melbourne, St Vincent's Hospital Melbourne, Fitzroy, Australia

[5]NICTA, Computer Vision Research Group, Canberra, Australia

[6]National Institute for Mental Health Research, Australian National University, Canberra, Australia

[7]Centre for Neural Engineering, University of Melbourne, National Information and Communications Technology Australia (NICTA), Ltd., Melbourne, Australia

ABSTRACT

Retinal visual prostheses ("bionic eyes") have the potential to restore vision to blind or profoundly vision-impaired patients. The medical bionic technology used to design, manufacture and implant such prostheses is still in its relative infancy, with various technologies and surgical approaches being evaluated. We hypothesized that a suprachoroidal implant location (between the sclera and choroid of the eye) would provide significant surgical and safety benefits for patients, allowing them to maintain preoperative residual vision as well as gaining prosthetic vision input from the device. This report details the first-in-human Phase 1 trial to investigate the use of retinal implants in the suprachoroidal space in three human subjects with end-stage retinitis pigmentosa. The success of the suprachoroidal surgical approach and its associated safety benefits, coupled with twelve-month post-operative efficacy data, holds promise for the field of vision restoration.

INTRODUCTION

Blindness is one of the most feared and debilitating physical disabilities [1], with the World Health Organization estimating that 285 million people worldwide have vision impairment and 39 million people are blind [2]. Recent technological advances have allowed the development of visual prostheses, or "bionic eyes", which use implanted electrodes within the visual pathway to restore rudimentary vision to those with severe vision loss [3], [4]. The success of these technologies has been such that two devices are now commercially available, with European CE mark approval being granted to the Argus II device (Second Sight Medical Products, USA) in 2011 followed by FDA approval in early 2013, and the Alpha IMS device (Retina Implant AG, Germany) gaining CE mark approval in mid-2013. These advances have provided proof of concept for the field of vision restoration, and brought hope to millions of people who have lost their sight from disease or trauma.

Visual prostheses work by using implanted electrodes to stimulate neurons in the visual pathway with electrical current via an implanted stimulator [5], [6] or independent microphotodiode-amplifier-electrode elements [7], [8]. The majority of visual prostheses developed to date use electrodes implanted within or near the retina to directly stimulate the inner retinal neurons, bypassing the non-functioning or dead photoreceptors, and utilizing the intact posterior visual pathway to transmit signals to the visual cortex. Thus, retinal prostheses require a relatively intact inner retinal layer to be beneficial, such as seen in retinal degenerative diseases like retinitis pigmentosa (RP) [9]. RP accounts for over 1.5 million cases of blindness and is the most common form of inherited blindness [10]. RP causes degeneration of vision over varying time periods and primarily targets the photoreceptor cells in the outer retina, with anatomical and retinal imaging studies showing that the inner retinal neurons are relatively preserved in these people [9], [11]–[13].

Previous work has shown that retinal prostheses in the epiretinal, subretinal and intra-scleral locations (Fig. 1) can provide patients with perceptions of light and basic object recognition, with some recipients even able to read large print [7], [8], [14]–[19]. However, there have been some challenges with previous approaches including issues with stability of the implant, conjunctival and scleral erosions, retinal detachment, hypotony and endophthalmitis [15], [20], [21]. Many of these complications are related to the device location and required implantation procedures, which include vitrectomy (removal of the gel-like vitreous humor from within the eye) and highly technical surgical techniques. Such adverse events can lead to poor vision outcomes, loss of residual vision or the need for revision/removal surgery.

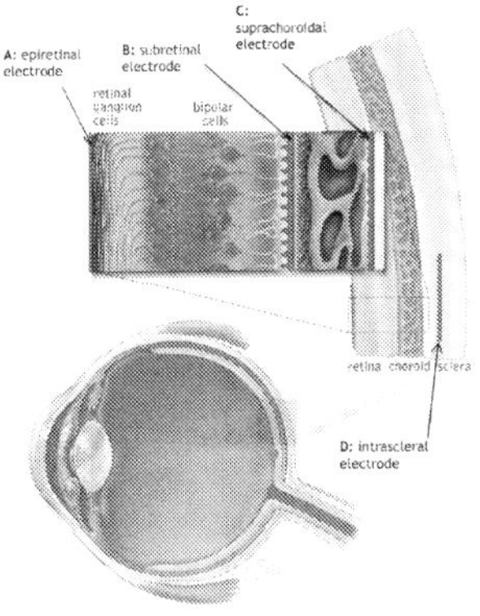

Figure 1. Potential anatomical locations for retinal prosthesis implantation. To date, clinical trials have been performed with devices in the A: epiretinal position [15], B: subretinal space [7] and D: intrascleral space [19]. Image modified with permission from Bionic Vision Australia.

The aim of this study was to develop a less invasive, more anatomically stable implantation position, by placing the array between the firm fibrous sclera and the outer retina/choroid, in the suprachoroidal space (Fig. 1C).

One of the main benefits with this location is that the surgical placement is less technically challenging and does not breach the retinal tissue, negating the need for a vitrectomy or incisions into the retina itself. The main disadvantage with this location is that the electrodes are 250 to 400 μm further away from the target retinal ganglion cells than in an epiretinal implant [22]. However, our group has shown in a preclinical feline model that not only is the suprachoroidal surgical approach feasible, reproducible and safe [23], the prosthesis is biocompatible, mechanically stable [24] and the electrode array is efficacious, with the ability to produce cortical responses within acceptable stimulation safety parameters [25], [26]. There have also been human studies showing that intra-scleral electrodes, which are implanted even further from the ganglion cells, are able to safely evoke phosphenes in patients with end-stage retinal degeneration at levels of stimulation considered to be safe for platinum electrodes [16].

This report describes the first human phase 1 clinical trial (clinicaltrials.gov, NCT01603576), involving three participants implanted with a retinal prosthesis in the suprachoroidal space. This study was designed as a proof of concept, first-in-human trial of the suprachoroidal implant. To allow for maximum flexibility in device stimulation, a percutaneous connector was used to allow direct stimulation of the intraocular electrode array, with no implantable electronics used. The percutaneous connector has previously been used in cochlear implant studies [27] and requires similar surgical procedures as the implantation of a bone-anchored hearing aid. Full specifications of the device are detailed in the following Methods section.

MATERIALS AND METHODS

Patient Screening and Selection

After approval from the Human Ethics Committee of the Royal Victorian Eye & Ear Hospital (application # 11/1032H) and trial registration (www.clinicaltrials.gov, trial # NCT01603576), three subjects aged 49, 52 and

63 years were enrolled in the study. Subjects were identified through a screening process at the Centre for Eye Research Australia, where 95 people with RP participated in a series of research projects as part of Bionic Vision Australia's development of the bionic eye.

An extensive battery of screening tests was used to assess suitability for the study, including visual performance tests (visual acuity, movement discrimination, Goldmann visual fields), confirmation of diagnosis through medical reports and clinical examination, measurement of residual retinal function using advanced Fourier transform analysis of electroretinography [28], assessment of the retina and choroidal structure using optical coherence tomography [22], assessment of functional vision in the home environment by an orientation and mobility instructor and an optometrist, psychosocial screening tests (including assessments of quality of life [29]) and a comprehensive consultation with a rehabilitation psychologist.

Initially, patient informed consent was provided for an 18-month study, but due to the successful progress in the first year (reported in this paper) and the availability of new hardware to enhance patient testing, a 6-month extension was offered to all participants at the twelve-month mark. All participants reconsented to the extension, making the entire study two years in duration. The second half of the study, including results of mobility testing and device explantation, will be detailed in a follow-up report. Independent psychological support was provided at all stages of the study and, in particular, during the informed consent periods.

The inclusion/exclusion criteria for this study are outlined in Table 1.

Table 1. Inclusion/exclusion criteria.

Inclusion Criteria	Exclusion Criteria
Aged 18 years or older	Any co-existent ocular disease, with the exception of mild cataracts
Either gender	Inability to visualize the retina due to corneal or other ocular media opacities (corneal degenerations, dense cataracts, trauma, lid malposition)
A confirmed history of outer retinal degenerative disease such as retinitis pigmentosa or choroideremia	Any ocular condition that predisposes the subject to rubbing their eyes
Remaining visual acuity of bare light perception or less in both eyes	Cognitive deficiencies, including dementia or progressive neurological disease
Functional inner retina (ganglion cells and optic nerve), as shown by the ability to perceive light and/or a measurable corneal electrically evoked visual response	Psychiatric disorders, including depression, as diagnosed by a qualified psychologist
A history of at least 10 years of useful form vision in the worse seeing eye	Deafness or significant hearing loss, or the presence of a cochlear implant
Must be willing and able to comply with the testing and follow-up protocol demands (preferably residing within 1.5 hours of the investigational site)	Poor general health or pregnancy, which would exclude them from obtaining a general anesthetic

doi:10.1371/journal.pone.0115239.t001

The three selected subjects provided written consent after being provided with both audio and electronic versions of the consent information. The subjects were a 52 year old post-menopausal female with rod-cone dystrophy and approximately 20 years of light perception vision (P1), a 49 year old male with syndromic RP (Bardet-Beidl syndrome) and approximately 8 to 10 years of light perception vision (P2), and a 63 year old male with rod-cone dystrophy and approximately 20 years of light perception vision (P3). All subjects were guide dog users. P2 and P3 had significant nystagmus, which did affect the ability to obtain high quality ocular images. P3 also had controlled epilepsy, for which he was taking valproate sodium. He was closely monitored by a neurologist during the study, and there was no evidence that the surgery or device stimulation changed his condition in any way.

All three subjects were assessed on four separate occasions over a two-year period prior to recruitment in the study whilst participating in research studies examining the natural progression of retinitis pigmentosa. The repeated measures allowed us to ensure that their baseline level of vision was stable, as people with RP often experience fluctuations in their visual performance [30], [31], and it is imperative to obtain an accurate baseline assessment of visual function prior to any vision restoration treatment.

The Suprachoroidal Retinal Prosthesis Implant

The device used in this study consisted of an intraocular electrode array (Fig. 2a), composed of a silicone substrate with 33 platinum stimulating electrodes (30×600 μm diameter, 3×400 μm diameter) and 2 large return electrodes (2000 μm diameter). The electrodes were recessed 50 μm from the silicone surface. A third return electrode was implanted remotely subcutaneously behind the ear. The outer ring of stimulating electrodes was ganged together to allow investigation of the use of hexagonal stimulation [32], meaning that there were a total of 20 electrodes that could be individually stimulated (Fig. 2c). The array was 19 mm long and 8 mm wide.

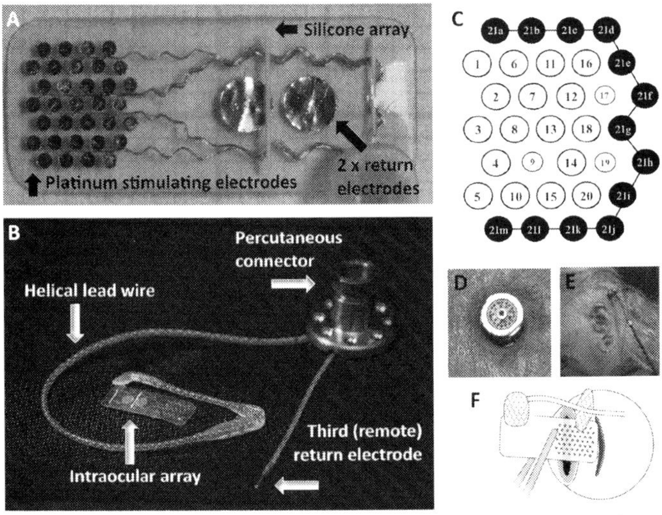

Figure 2. The intraocular electrode array of the suprachoroidal device (A) and the entire device (B), showing the array connected to the percutaneous connector via a helical lead wire. The electrodes on the intraocular array (C) were numbered for analysis, with the black electrodes (21a to 21m) being ganged to provide an external ring for common ground and hexagonal stimulation parameter testing. Note electrodes 9, 17 and 19 were smaller (400 µm vs. 600 µm). The percutaneous connector protruded through the skin behind the ear (D), allowing direct connection to the neurostimulator via a connecting lead (E). The scleral incision was made 9 mm to 10 mm posteriorly from the sclero-corneal limbus.

The intraocular array was connected by a helical lead wire to a titanium percutaneous connector (Fig. 2b), which was implanted behind the subject›s ear. The lead wire was 150 mm long and contained 23 individually insulated platinum/iridium wires. The 24th wire from the remote platinum return connected directly to the percutaneous connector.

The percutaneous connector has previously been used in cochlear implant studies [27] and required similar surgical procedures as the implantation of a bone-anchored hearing aid. The connector was anchored to the skull using titanium screws, and protruded through the skin behind the ear (Fig. 2d). The advantage with the percutaneous connector was that it enabled direct stimulation of the electrode array via a connecting lead (Fig. 2e), without the need for implantable electronics. This allowed unparalleled flexibility in the testing paradigms possible.

Suprachoroidal Surgical Technique

The surgical procedures for the implantation of the suprachoroidal retinal implant and the percutaneous connector were developed as an iterative process, incorporating both cadaver studies (feline and human) and acute and chronic (3 months) implantation in preclinical studies using a feline model [33]. An acute study was also completed in a human patient who was undergoing enucleation due to an unrelated ocular condition. This acute study enabled surgeons to trial the procedure with an inactive mock device prior to the commencement of the clinical trial. Full details of the surgical procedures can be found in a previous publication by Saunders et al [33], and are summarized here.

All three operations occurred at the Royal Victorian Eye and Ear Hospital, Victoria, Australia between May and August 2012, with vitreoretinal surgeons PA, JY and MM and ENT surgeon RB. General anaesthesia was administered and the area was prepared near the site of the percutaneous connector by shaving the hair and sterilizing the location with Betadine. The eye was also prepared with a Betadine wash. A dummy electrode array and percutaneous pedestal were used at this early stage of surgery to plan and mark the required incisions.

A curved scalp incision was used to expose the posterior temporalis muscle, followed by an incision through the muscle to expose a flat section of squamous temporal bone. The periosteum was dissected from the bony site to allow attachment of the percutaneous pedestal. A tunnel was created beneath the temporalis fascia between the percutaneous connector location and the lateral orbital rim.

A lateral canthotomy was performed and extended towards the ear and an incision made in the flap for the percutaneous portion of the device. After the lateral orbital margin was exposed, the periosteum was incised and a flap created. A lateral orbitotomy was created with 10 mm burrs below the zygomaticofrontal suture for lead stabilization. The electrode array was loaded into a purpose built trochar and then passed forward to the lateral canthotomy/lateral orbital margin. The percutaneous pedestal was secured with self-tapping screws and the external ground electrode was placed beneath the temporalis periosteum close to the pedestal.

The scleral incision was made behind the lateral rectus muscle, to use the muscle to cover the Dacron patch and reduce the risk of exposure.

As such, a temporal peritomy was performed and the lateral rectus muscle identified. Lateral rectus was disinserted and tied with 6/0 vicryl (Ethicon). The scleral wound position was marked with diathermy (as decided from preoperative axial length measurements, ultrasonography and magnetic resonance imaging). The location of the scleral wound is demonstrated in Fig. 2f. Due to the axial length, globe dimensions and lateral rectus muscle insertion location, P3 had a more posterior incision location (10 mm posterior to the sclera-corneal limbus) than P1 and P2 (both 9 mm posteriorly from the limbus). This meant that the electrode array was placed more temporal from the fovea for P3 than the other subjects. A full thickness scleral wound was made with a 15 degree blade (Alcon) and the suprachoroidal space initially dissected with a crescent blade (Alcon), then with a lens glide (bvi Visitec 581001).

The electrode array was inserted into suprachoroidal space and the superior wound edge was extended posteriorly in an L shape to allow the lead to exit and the Dacron patch to sit flat on the globe. This L shaped extension allowed the device position to be optimized during insertion. The wound was closed with 8/0 nylon (Ethicon) and the Dacron patch sutured to the sclera with 8/0 nylon.

After the scleral wound was closed, the lateral rectus muscle was reattached. The device position was assessed with indirect ophthalmoscopy. Due to the subretinal hemorrhage noted three days postoperatively in P1, we attempted to minimize eye movements in P2 and P3 in the immediate post-operative recovery period by injecting Botulinum toxin Type A (Allergan Australia, NSW, Australia) into the four extraocular rectus muscles. This did not change the outcome for P2 and P3, with both also experiencing a mild to moderate subretinal bleed in the 3–4 day postoperative window.

At the conclusion of the surgery, the conjunctiva was closed with 8/0 vicryl. The lead was fixated in an orbitotomy drilled in the frontal process of the zygomatic bone, and held in place by a specifically designed silicone grommet. The periosteum was closed over this with 6/0 vicryl sutures. The subcutaneous tissue wound was closed with 6/0 vicryl and the skin sutured with 5/0 nylon or staples.

An injection of subtenon's local anaesthesia was used for short-term pain relief and chloramphenicol antibiotic ointment was applied to the eye. Finally, a pressure dressing was applied to the area surrounding the percutaneous connector. This was removed the following day.

Subjects remained in hospital for four to five days of postoperative care before being discharged. They were given postoperative topical steroid (predneferin forte) and antibiotic (chloramphenicol) eye drops, and analgesia as required. After discharge, subjects were reviewed weekly and monitored for adverse events, post-operative safety and clinical efficacy as described below.

Assessment of Surgical Safety and Efficacy

Intraoperative safety was defined as the number of adverse events during the surgical procedure. Surgical efficacy was defined as the proximity of the electrode array to the fovea following implantation, and the mechanical stability of the device in the suprachoroidal space. Both of these assessments were made from quantitative measurements on ocular imaging modalities (fundus photographs and optical coherence tomography).

Assessment of Post-Operative Safety

Post-operative safety was assessed by the number of serious device related adverse events following the implantation. We defined a serious adverse event as one that required altered or increased medical management, such as surgery or hospitalization. To be considered as "device-related" there needed to be a direct or indirect causal link between the device, implantation surgery or study protocols and the adverse outcome.

Assessment of Device and Clinical Efficacy

Device efficacy was determined by the number of electrodes that remained connected and viable after implantation (as indicated by electrode impedances) and the ability for subjects to perceive repeatable phosphenes when the device was stimulated.

Electrode impedance was measured at the end of the cathodic (first) phase of a biphasic current pulse and defined as the quotient of dividing the peak measured voltage by the supplied current [34]. Individual electrode impedances were measured using clinical software running on a laptop PC. Measurements were made intraoperatively to ensure that the implant was functional before closing the surgical wounds. After surgical recovery, impedance measurements were made using the same laptop-based system approximately fortnightly for the duration of the study.

Visual function was assessed via two tests – the Basic Assessment of Light and Motion (BaLM) test [35] and the Landolt C optotype recognition subtest from the Freiburg Acuity and Contrast Test (FrACT) [36], both of which were presented to subjects in a darkened room (108–114 lux) using a 30-inch computer monitor placed at 57 cm viewing distance. The BaLM test was completed with all subjects, but only P1 completed the Landolt C testing protocol in the initial 12-month post-operative period. Testing incorporated a head-mounted video camera with a manufacturer stated field-of-view of 67°×50.25° (Arrington Research Inc., Scottsdale AZ, USA) and a pixel dimension of 320 by 240 pixels. Within the implant, the 20 stimulating electrodes are arranged in a staggered grid measuring 3.5 mm×3.46 mm, corresponding to a visual field projection on the retina of 12.4°×12.2° [37].

The BaLM test involves detection of a light wedge in one of four quadrants, and assesses the ability of the device to improve light localization skills [35]. Given the response options were four alternative forced choices (4AFC), the chance rate was 25% and the criterion cutoff for success set at 62.5%. The BaLM was completed with the device off (control) and device on. The device on setting used a vision processing algorithm called Lanczos2 filtering, which is reported in detail elsewhere [38], [39]. In short, the Lanczos2 filter is a reconstruction filter that is used in downsampling and translating the high resolution image from the camera to electrode stimulation parameters. The filter ensures artefacts from such downsampling do not appear, such as a flickering which may result from making small head movements with the camera viewing fine detail. This makes objects appear more consistent in their appearance. The percentage of accurate responses in the BaLM test was determined for each participant with device on and device off. Comparisons between participants and device setting for percentage of accurate responses were calculated using χ^2 statistics. Binomial distributions were calculated to determine whether the accuracy rates were significantly better than chance (i.e., 25%). Windows SPSS v22 (IBM Corporation, Somers, NY) was used for all statistical analyses.

The Landolt C optotype recognition test used was a subtest from the FrACT [36], which is freely available online (http://michaelbach.de/fract/index.html). This task requires subjects to identify the orientation of a letter C, and can be used to estimate visual acuity. It was not possible to present appropriately sized optotypes on our 30-inch computer monitor, so a region-of-interest was defined within the camera video image to be 18×18 pixels, corresponding to a reduced visual field of 3.77°×3.77° (a 'zoom' of 3.26). Measured thresholds have this correction factor applied. The theoretical acuity for this device was assumed to be the minimum grating distance that can be represented in four orientations. This is limited by the 1 mm horizontal electrode spacing, and corresponds to a grating of 0.141 cycles-per-degree, or 2.33 logMAR (20/4242). We only completed the Landolt C test on P1, and she was tested over 32 sessions. In total, this testing involved 19 trials with the device on (using Lanczos2 vision processing filter), and 2 trials with the device off. Due to the obvious floor effect with device off, there was no need to repeat this further. The patient wore a head-mounted camera that could be focused on different regions of the screen by adjusting her head azimuth or elevation. She was trained to find the center of each optotype by first judging it's width and height. From this central fixation point, the patient explored regions in each of the four possible directions, returning to the center each time and repeating until ready to respond. Descriptive analyses and the non-parametric Wilcoxon Rank Sum test were performed on the Landolt C data using Windows SPSS v22 (IBM Corporation, Somers, NY).

RESULTS

Three subjects with profound vision loss from retinitis pigmentosa consented for implantation; two males aged 49 and 63 years and a 52-year-old post-menopausal female. They were implanted with the suprachoroidal retinal prosthesis and percutaneous connector in mid-2012. Follow-up data were available for twelve months for each subject. Details of the clinical characteristics of subjects, electrode design/fabrication and surgical procedures are described in full in the methods.

An optical coherence tomography (OCT) cross-sectional scan of the device *in vivo* is shown in Fig. 3. Due to the fact that the choroid becomes thinner in late stages of RP [22], the electrode array is easily visualized between the choroid and sclera.

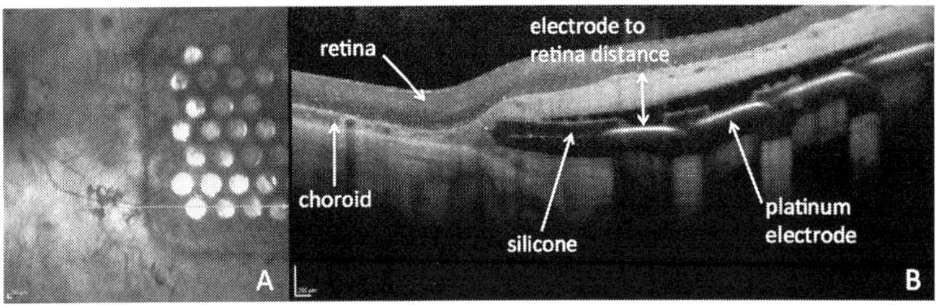

Figure 3. OCT scan of the electrode array in situ, taken 2 months postoperatively in Patient 1. The horizontal arrow on the infrared image indicates the direction of the OCT scan (A). The cross-sectional OCT image (B) shows the silicone and platinum electrode components of the array, the retina structure and choroidal vasculature, and the electrode to retina distance used for analysis (double-headed arrow). Scale bars = 200 μm.

Surgical Safety and Efficacy

Implantation surgeries were performed under general anaesthesia and took three to four hours, with the duration decreasing between surgeries due to experience. The surgeries progressed without intraoperative complications. Due to the choroidal atrophy, the electrode array was easily sighted using conventional indirect ophthalmoscopy techniques during surgery, which allowed the device to be placed in the optimal position underneath the macula in all three cases. No intraocular or suprachoroidal hemorrhage was noted during the surgery.

Postoperative Clinical Results: Retinal Health

The postoperative recovery in all three subjects was similar to other major eye surgeries with respect to reported pain levels and the use of analgesia. Patient 1 (P1) reported more pain around the percutaneous plug implantation site (requiring narcotic analgesia with morphine 2.5 mg for the first two days) than P2 and P3, both of whom only required simple analgesia (oral paracetamol 1 g for the first two days). The level of intraocular inflammation was mild and similar to that seen in common procedures such as a retinal detachment repair, and treated with topical steroid (Prednefrin Forte, Allergan Australia, NSW, Australia) and antibiotic (Chlorsig, Aspen Pharma, NSW Australia) eye drops four times daily. There was no change in the intraocular pressure of the three participants over the duration of the trial, in either the implanted or control eye. One patient (P1) experienced mild to moderate limitation in the abduction of the implanted eye following surgery, but this function improved over the duration of the study and did not cause the patient any functional difficulty nor cosmetic concern.

In all three subjects, a combined subretinal and suprachoroidal hemorrhage formed three to four days postoperatively, which resolved without any ongoing complications in P1 and P3. In P2, who experienced a larger hemorrhage than the others, a fibrotic tissue reaction remained at the temporal end of the implant following hemorrhage resolution, but this did not affect device efficacy nor cause complications such as retinal detachment. None of the subjects were taking anti-coagulant medication during this study. We did attempt to reduce eye movements in P2 and P3 by using intraoperative botulinum toxin injections into the rectus muscles and medications to reduce rapid eye movements during sleep, but the reduced eye movements did not prevent this hemorrhage from forming. Complete resolution of the hemorrhage took 55 days in P1, 101 days in P2 and 13 days in P3. Fundus photos showing the time course of the haemorrhage resolution for P1 are shown in Fig. 4. None of the subjects required any intervention during the resolution of the hemorrhage. Testing of the device did not commence until there was significant resolution of the blood, to allow clear visualization of the retina prior to any stimulation.

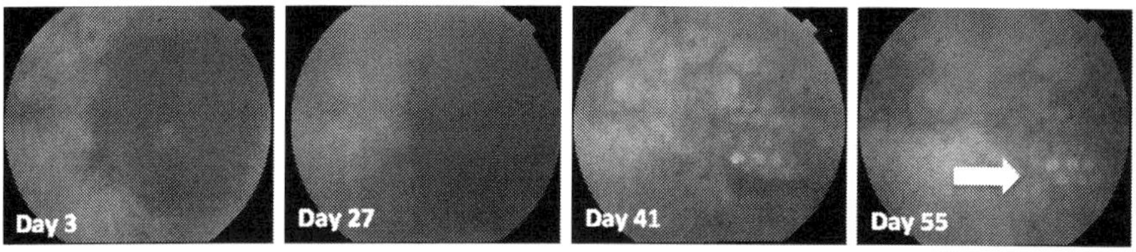

Figure 4. Time course of subretinal hemorrhage in P1, as documented with retinal fundus photography. Complete resolution of the hemorrhage occurred in this subject by 55 days post-operatively. Note the electrode array with individual electrodes can be seen more clearly over time as the blood clears in the temporal retina (arrow).

Longitudinal OCT measurements taken over the 12 month post-operative period demonstrated that there was no change in the retinal thickness above the electrode array, suggesting that no significant retinal oedema or atrophy had occurred after the implantation of the array (Fig. 5). Measurements of retinal thickness were taken directly above each of the 33 stimulating platinum electrodes, and the results shown reflect the data from all points. Note that as P2 and P3 had significant nystagmus, there was greater variability in the measurements due to decreased image quality and the fact that fewer locations were able to be sampled. Due to persistence of the subretinal and suprachoroidal hemorrhage in P2, retinal thickness measurements were not possible until Day 143.

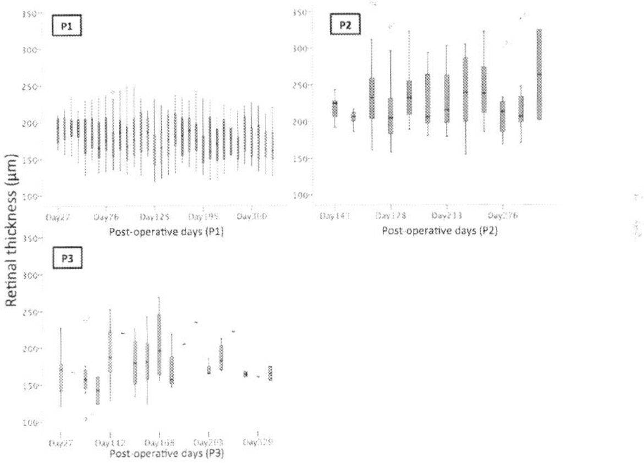

Figure 5. Retinal thickness measurements over time, showing no observable change in the maximum retinal thickness above the electrode array in the initial twelve months in all three patients. Each boxplot includes the maximum (upper whisker, excluding outliers), upper quartile (top of box), median (horizontal line in box), lower quartile (bottom of box) and minimum values (lower whisker). Open circles are outliers. The numbers represent the electrode location (see Fig. 2), and the horizontal lines on the graph of P3 represent single data points.

Postoperative Clinical Results: Summary of Device-Related Serious Adverse Events

Subjects returned for post-operative assessment on a weekly basis during the entire study. The only device-related serious adverse event was a staphylococcus aureus infection around the percutaneous connector in P2 at day 59, which required a three-day hospital admission. The infection was successfully treated with a combination of intravenous vancomycin and flucloxacillin, and oral rifampicin and fusidic acid. The infection did not necessitate removal of the implant, nor did it have any effect on the stimulation or psychophysics testing. No device-related serious adverse events occurred with involvement of the intraocular electrode array.

Postoperative Clinical Results: Device Stability

Intraocular array stability was monitored using fundus photography, infrared (IR) imaging and OCT. Measurements were taken of the position of the electrode array with respect to the optic nerve and vascular landmarks, and showed there was no lateral movement of the array with time. This stability can be seen in the time series of IR images shown in Fig. 6. Over time the individual electrodes became more difficult to visualize in P2 and P3.

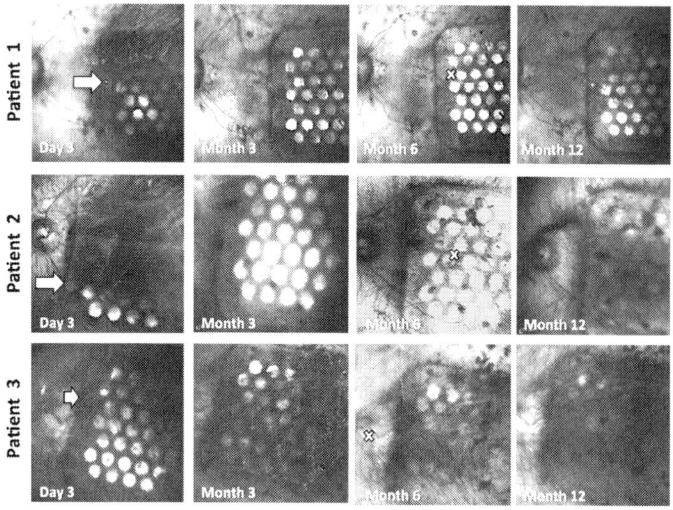

Figure 6. Time series of IR images of the electrode array in all subjects, showing no significant lateral movement of the array over a twelve-month period. The leading edge of the array is marked with a white arrow in the 3-day images. The position of the fovea is marked with a white cross in the 6-month images.

The distance between the retina and the electrode array was measured from the cross-sectional OCT scans, and was taken from the anterior surface of the platinum electrode disc to the posterior surface of the retina (the retinal pigment epithelium, or RPE), as shown in Fig. 3.

After the subretinal hemorrhage resolved, the electrode to retina distance has remained relatively constant for P1, but has increased for

P2 and P3, Fig. 7. This increase in separation is visible on OCT scans, which show a hyper-reflective band between the choroid and the retina. The nature of this band is still under investigation.

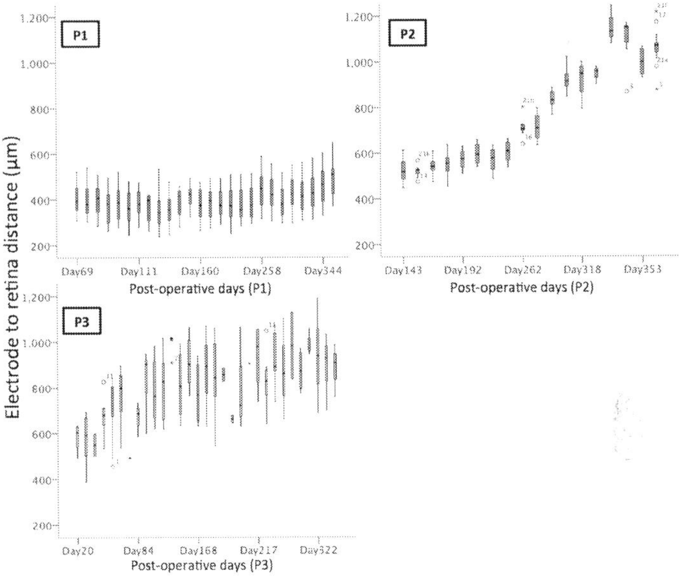

Figure 7. Distance between the electrodes and retina over time. This distance was relatively constant in P1, but increased up to two-fold in P2 and P3 over time. Note significant nystagmus in P3 made the measurements difficult, leading to greater variation in the values recorded. Outliers are identified by open circles, and the numbers represent the electrode location (see Fig. 2).

Stability of the titanium percutaneous connector and helical lead wire was monitored using monthly X-ray images. Comparison of the array and lead wire position was made with respect to anatomical landmarks, in three primary positions of gaze (straight ahead, left and right gaze). This also provided information about physical integrity of the lead wire, which showed no signs of stress or lead wire breakage. Over this initial twelve-month period, there was no significant movement of the intraocular array, lead wire or percutaneous connector on the X-ray images (Fig. 8). Note that the position of the zygomatic notch varies between subjects, hence there is variability in the vertical position of the lead entry point in the three subjects. The lead wire formed a broad superior loop in P1, and followed the temporal wall more closely in P2 and P3.

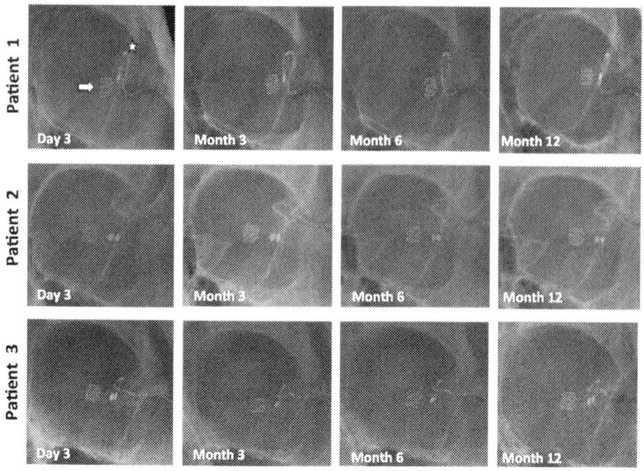

Figure 8. Stability of the intraocular array (arrow) and helical lead wire (star) over the initial 12 months of implantation, as documented by X-ray images. Additional scans were taken to monitor the percutaneous connector (not shown), which also stayed stable over this time period.

Postoperative Clinical Results: Device Integrity

Electrode channel continuity was verified by measuring intraoperative and post-operative electrode impedances. We measured impedance by applying a small test current pulse and measuring the electrode voltage. The resultant voltage was divided by the applied current to calculate impedance [40]. In all subjects, 100% of the electrodes within the implant remained intact following surgery and remained connected over the twelve-month period.

Electrode impedance data were averaged monthly, from the date of implantation onwards, for each subject's array. The distribution of these data are plotted as box plots (Fig. 9A–C). The mean impedances for the 600 μm electrodes in P1 and P2 ranged over time from 9.7 to 14.5 kΩ and for P3 the mean impedance ranged over time from 3.6 to 13.3 kΩ. Impedance measurements were stable over the passive implantation and psychophysics testing periods in P1 & P2, but decreased significantly in P3 (Fig. 9C). This decrease in impedance suggests a change in the electrode- tissue interface or in the tissue surrounding the array. Further investigation continues to ascertain the exact mechanism of the decreased impedance for P3.

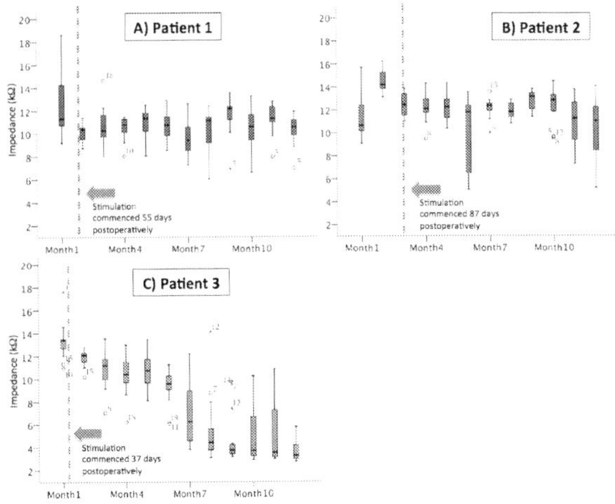

Figure 9. Impedances for the 600 **μm** platinum electrodes over time in the three subjects.

Impedances were measured with charge-balanced biphasic current pulses (pulse phase width: 25 μs; amplitude: 75 μA). The dotted lines represent the date of first stimulation. In P1 & P2, the impedances were stable over the implantation and stimulation period. Impedances measured in P3 decreased over the course of the implantation period. Outliers are identified by open circles, and the numbers represent the electrode location (see Fig. 2).

The range of measured impedances, particularly for P3, included some markedly low values. However, follow-up device integrity testing (electrode voltage tomography) on all three subjects did not reveal any evidence of short circuits in any of the electrode channels, suggesting that these low impedance values were genuine.

Postoperative Clinical Results: Device Efficacy (Phosphene Perception)

Following post-operative recovery and allowing time for almost complete resolution of the hemorrhage over the electrodes, stimulation of the electrode array commenced for weekly psychophysics sessions of between 2 and 5 hours. The first stimulation session was held 55 days postoperatively for P1, 87 days for P2 and at 37 days for P3. Stimulation was delivered using a custom-built neurostimulator ("neuroBi"), which allowed direct stimulation of the individual electrodes via connection with the percutaneous connector [27] and is designed to allow flexible configuration of testing parameters [41]. The stimulator delivered charge-balanced biphasic current pulses with pulse widths ranging from 100–1000 μs per phase. The electrodes were capacitively coupled and shorted between current pulses to remove any potentially damaging residual charge [42]. Unless otherwise stated, the electrodes were stimulated in a monopolar electrode configuration using one of the 2 mm diameter platinum electrodes as the return (see Fig. 2).

Reliable phosphene percepts were elicited in all three subjects. To ensure that overstimulation above safe charge and charge density levels did not occur, the charge per phase on each electrode was capped at 447nC for the 600 μm electrodes and 298nC for the 400 μm electrodes (which equates to an upper limit of charge density of 158 $\mu C/cm^2$ and 237 $\mu C/cm^2$). These values were determined using the Shannon model of safe levels for electrical stimulation, using a conservative k value of 1.85, as defined in Merrill et al [43]. Within this safe charge limit, percepts were able to be obtained on all 20 electrodes for P1 and 18/20 electrodes for P2 using monopolar stimulation at a rate of 50 pulses per second (pps). By increasing the stimulation rate to 500pps, percepts could be obtained from all 20 electrodes in P2.

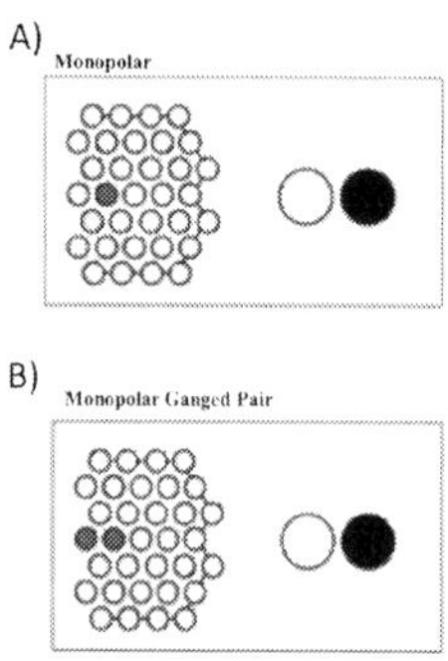

C)	Patient 1	Patient 2	Patient 3
Low rate stimulation on single electrodes (Monopolar, 500μs PW, 500μs IPG, 50pps, anodic first)	20 / 20 electrodes tested	18 / 20 electrodes tested	Not tested
High rate stimulation on single electrodes (Monopolar, 500μs PW, 500μs IPG, 500pps P1= cathodic first, P2 & P3= anodic first)	6/6 electrodes tested	20 / 20 electrodes tested	12/14 electrodes tested (limited dynamic range)
High rate stimulation on ganged electrode pairs (Monopolar ganged pairs, 500μs PW, 500μs IPG, 500pps, anodic first)	Not tested	Not tested	9 / 9 ganged pairs

Figure 10. Different monopolar electrode stimulations used in the psychophysical testing of the implanted subjects (A & B), where the active electrodes are shown in red and the return electrodes in black. Table (C) shows the number of electrodes that were capable of eliciting a visual percept using the given stimulation parameters in each subject. PW= phase width, IPG= interphase gap, pps= pulses per second. Stimulus duration was 2 seconds in all cases.

In P3, it was possible to elicit phosphenes using individual electrodes, however charge levels approaching the safe limit were required. Stimulation was subsequently performed using ganged pairs of adjacent electrodes as they provided greater dynamic range, allowing brighter percepts to be produced (Fig. 10B). Using ganged pairs increased the maximum safe charge per phase due to the larger total electrode area, thus enabling a 100% utilization rate for all ganged pairs tested with P3.

Phosphene appearance was found to be varied depending on subject, electrode position and stimulation parameters, but were controllable (in terms of size and, brightness and complexity), retinotopically placed and locatable in the visual field by the subject [44].

Visual Function Results: Light Localization

In all subjects, the light localization sub-task of the Basic Assessment of Light and Motion (BaLM) test [35] was completed to assess the device's ability to improve light localization. In this test, the subject needs to identify the orientation of a wedge of light on a black computer screen (up/down/left/right), and it is a four alternative forced choice paradigm with a chance rate of 25%.

For the device on setting (using a Lanczos2 filter), all three participants performed significantly better than the 25% chance level. achieving 97.50% (p<.0001), 71.43% (p<.0001), and 66.66% (p<.0001) correct response rates for P1, P2 and P3 respectively. All of the participants performed at a level at or below chance when the vision processing system was switched off; Table 2.

Table 2. Results of the BaLM light localization task, showing that using the device (with a Lanczos2 vision-processing filter) gave significant visual improvement compared with device off, p<0.0001 for all three subjects.

	Device On			Device Off			
Subject	Number of trials	Number correct	Percentage correct	Number of trials	Number correct	Percentage correct	Difference between device on and device off
P1	40	39	97.5%	72	20	27.78%	P<0.0001
P2	56	40	71.43%	40	10	25%	P<0.0001
P3	48	32	66.66%	40	10	25%	P<0.0001

doi:10.1371/journal.pone.0115239.t002

Visual Function Results: Optotype Acuity Estimates

As P1 was implanted first, we have results for a longer post-implant time than for the other two participants. This enabled us to begin to measure optotype recognition using the Landolt-C task in the FrACT visual test battery [36]. For the device on setting (using the Lanczos2 filter), visual acuity was estimated to be 2.62 logMAR on average (range 2.35 to 3.02) corresponding to 20/8397 (range 20/4451 to 20/21059) over 19 sessions (Fig. 11). This compares favorably to the calculated limit of 2.33 logMAR (20/4242) for a 1 mm electrode pitch [37]. For the device off setting, P1 was unable to see any of the Landolt C optotypes at all. Due to the limit of the software (floor effect) we were unable to estimate any visual acuity lower than 3.24 logMAR. The average response time per optotype presentation was 59 seconds for device on (range 43 to 80 seconds) and 11 seconds for device off.

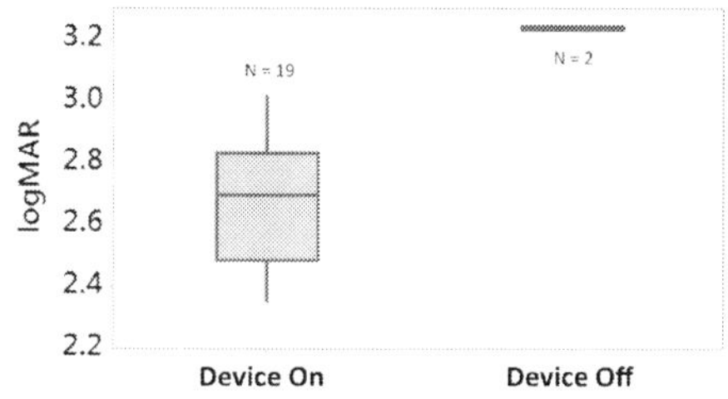

Figure 11. Optotype recognition results for P1 using the Landolt-C test, which gives an indication of acuity (but should not be directly interpreted as standard visual acuity). Whiskers extend to minimum and maximum recorded thresholds and the box extents show inter-quartile range. The Lanczos2 vision processing performed significantly better than System Off (P=.01). Note that the visual acuity exceeded the 3.24 logMAR software limit in all trials with device off (floor effect).

The non-parametric Wilcoxon Rank-Sum test was employed to determine whether there were differences between Lanczos2 vision processing and System Off in performance for visual acuity (logMAR) using Landolt C optotypes. Lanczos2 (N=19, *Median*=2.70) vision processing performed significantly better than System Off (N=2, *Median*=3.235) for P1 with $z=-2.280$, p=0.010; Fig. 11].

DISCUSSION

This world-first Phase 1 trial of a retinal implant in the suprachoroidal space has shown that the novel surgical procedure is safe, the position is stable and the device was able to provide visual percepts in all three subjects. The electrode array was well tolerated by the eye and all electrodes on the three arrays remained functional over the 12-month monitoring period.

The main advantage with the suprachoroidal surgical approach is the ease and safety of the surgical procedure, as well as the proven device efficacy in all three of our subjects. It is remarkable that all electrodes remained functional for our subjects for the 12-month period, as this has not been the case with previous devices. *For example*, a report to the US Food and Drug Administration in September 2012 reported that in the Argus II 60-electrode device clinical trial, an average of 55.5±3.6 electrodes were functioning directly after implant, with another 1.2±1.8 electrodes automatically disabled during stimulation due to high impedances [20].

The implantation surgery for each device took less than four hours, and there were no intraoperative adverse events in any subject. It is anticipated that the surgical procedure used for the implantation of these suprachoroidal devices could be undertaken by the majority of ophthalmic surgeons, which to date has not been the case for the other retinal approaches. It is also anticipated that this location may allow for simpler removal and replacement with upgraded devices in the future. This has been demonstrated in preclinical feline models [45] but not yet in humans.

The devices have been mechanically stable over the initial postoperative twelve months, with no lateral movement (with respect to the retinal vasculature landmarks) nor signs of extrusion. The one serious device-related adverse event that did develop in this initial twelve-month period was an infection of the percutaneous connector, which was an anticipated risk (having been noted in previous trials using percutaneous connectors for cochlear implant research) and was successfully treated with intravenous antibiotics with no damage to the device, nor need for explantation. The use of a percutaneous connector allowed direct and flexible stimulation of the intraocular electrode array, but required a metallic foreign body to protrude from the skull for the duration of the study. In subsequent studies it is planned that the percutaneous connector will be replaced with a fully implantable and wireless stimulator system, hence eliminating the risk of such infections in future iterations.

The other adverse event that was noted was that all three subjects had a combined subretinal and suprachoroidal hemorrhage three to four days post-operatively, which resolved without complication in two patients. In P2, a small fibrovascular scar remained at the edge of the implant following haemorrhage resolution, which has been stable for 12 months. After noting this in our first subject, the study protocol was changed to include intraoperative botulinum toxin injection and post-operative oral medication to reduce rapid eye movement during sleep, in the hope that a more stable eye during the initial healing period would prevent the formation of this hemorrhage. However, this was not the case, and the exact mechanism of the hemorrhage remains unclear.

Due to the suprachoroidal implant location, we were concerned about the risk of damaging the long posterior ciliary neurovascular bundle during implantation, as this bundle pierces the sclera and crosses the suprachoroidal space [46]. We were prepared for the risk of contacting this bundle with a surgical contingency plan, which included methods of diathermy to stem any bleeding, as outlined previously [33]. However, this was not required in any of the patients. Due to the fact that the postoperative bleeding originated in the anterior choroid (not in the suprachoroidal space) and occurred in the postoperative period, we do not believe that this was the cause of the observed hemorrhages. However, the presence of the long posterior ciliary neurovascular bundle does need to be considered during suprachoroidal implant surgeries.

As can be seen in the IR images (Fig. 6), the electrodes have become more difficult to visualize with time, especially in P3. This is correlated with an increase in the electrode to retina distance in P3. This may be due to a change in the tissue-electrode interface, but due to limitations in ocular imaging techniques associated with the subject's nystagmus, we cannot conclusively determine whether this area is fibrous tissue, fluid or some other material. Ongoing investigations are continuing, and it is likely that we will need to wait until explantation of the devices to gain further information. This outcome was not noted in our preclinical feline models.

Intra-operative and post-operative impedance testing demonstrated ongoing electrode continuity over a twelve-month follow-up period. These results suggest a robust electrode array, lead system and connector. This is a significant finding, as it is very common for some electrodes and even the lead wires to break during implantation surgery in other locations. In addition to demonstrating surgical safety, the fact that all electrodes within the clinical arrays were viable after twelve months implantation shows the mechanical robustness and stability of the suprachoroidal position and associated lead routing path.

Another interesting finding has been the decrease in impedances in P3 over time. Whilst P1 and P2 maintained a mean impedance of 9.7–14.5 kΩ over the study period, impedances for P3 dropped from an initial mean measure of 13.3 kΩ to a mean of 3.6 kΩ by twelve months. A drop in impedance suggests a change in the electrode-tissue interface or the tissue surrounding the array, but the exact cause for these changes are not known, with further investigations underway to ascertain the potential mechanism.

The impedance results are comparable to those derived from previous studies in our feline model, using a similar suprachoroidal position for the electrode array with platinum electrodes of the same dimensions [40]. These preclinical studies showed stable mean impedances in the range of 11–15 kΩ over a 5 month implantation period with approximately three months of continuous suprathreshold electrical stimulation in the suprachoroidal space. The similarity between impedances measured clinically and preclinically adds support to the veracity of the feline suprachoroidal model published previously by our group, which will be used for ongoing development of future devices.

We were able to generate visual percepts in all three subjects within a safe charge limit. It was necessary to use ganged pairs of electrodes to gain useful percepts with a satisfactory dynamic range for P3, which may be due to the increased distance between the array and the retina, or due to the array being further from the macula than in the other two subjects (see Fig. 6). Unlike the other two participants, P3 is also on anti-epileptic medication (valproate sodium 100 mg), which potentially may have a role determining the parameters needed to record a percept. Further details on the factors that affect phosphene perception in this cohort are detailed in a separate publication by Shivdasani et al [47].

As this was an initial proof of concept study, with only three participants were recruited, it was not possible to draw inferences about the effect of gender, age, surgical ancillary treatment (i.e. the use of Botox injections during implantation) or systemic medications on the outcomes. It is certainly of interest that the results varied between our three participants, and this gives support to our belief that device fitting and outcome assessment for bionic eyes will remain a time-consuming and individual process.

Device efficacy was also shown using a test of light localization (BaLM test) and a Landolt-C optotype recognition test (FrACT test). Light localization is a simpler visual function than optotype recognition, but does not enable an estimate of visual acuity. We demonstrated that all subjects showed improved light localization with the device on (all passing the test's "pass" criterion of 62.5%), with performance with device off below the level of chance. We found that the Landolt-C test required significant orientation and habituation, which was not possible in the 12-month postoperative time frame for P2 and P3. As P1 was implanted first, we were able to estimate her optotype acuity using this test, and showed that it was significantly improved with device on (best logMAR measured=2.35) vs. device off (logMAR worse than 3.24, reached the floor effect of the test). Previously reported visual acuity from other groups have been up to 20/1262 (logMAR 1.8) for the Argus II epiretinal device (60 electrodes) using grating acuity tests [15] and 20/546 (logMAR 1.43) for the Retina Implant AG Alpha IMS device (1500 independent microphotodiode-amplifier-electrode elements) using the Landolt C optotype acuity test [48].

It should be noted that the use of Landolt C optotype recognition tests in prosthetic vision gives estimations of acuity only, and cannot be directly compared with acuity tests in normally sighted individuals. Subjects with a visual prosthesis will use different techniques to assess the optotypes, including head scanning and learnt interpretation of the phosphenes. Thus the visual acuity reported using visual prostheses is likely to be overestimated. Future studies will be vital to obtain more detailed information regarding the potential visual acuity of the suprachoroidal devices in a greater number of participants, and this estimated acuity value should be interpreted with caution, and whilst remembering that our prototype device only had 20 stimulating electrodes.

This study has shown the suprachoroidal anatomical position is a viable, minimally invasive and relative straightforward location for an electrode array, with successful stimulation of visual percepts and excellent lateral device stability in all subjects. The device efficacy is expected to improve with technological advances in the next iteration, and also with the ability for patients to be trained on use in their home environment in future trials (which was not possible in this study). We will continue to investigate the presence of the band between the device and choroid, including using histopathological assessment techniques on explanted devices at the end of this study (which will cease in late 2014). The inherent advantages of the suprachoroidal position, including ease of surgery (no vitrectomy or retinal incision required), robustness and stability of the electrodes, and minimal implant-related intraocular serious adverse events suggest that this approach has promise in the field of retinal prosthesis research.

ACKNOWLEDGMENTS

The Bionic Vision Australia Consortium consists of five member organizations (Centre for Eye Research Australia, Bionics Institute, NICTA, University of Melbourne and University of New South Wales) and three partner organizations (The Royal Victorian Eye and Ear Hospital, National Vision Research Institute of Australia and the University of Western Sydney). For this publication, the consortium members consist of Prof Nigel Lovell, Prof Gregg Suaning, Prof Hugh McDermott, Prof Peter Seligman, Ms Adele Scott, Dr Wilson Heriot, Mr Owen Burns, Mrs Alexia Saunders, Ms Michelle McPhedran, Mr Ronald Leung, Dr James Fallon, Ms Ceara McGowan, Mr Kyle Slater, Dr Thushara Perera and Ms Tamara Brawn.

We thank Professor Gregory Murphy from the School of Public Health at Latrobe University for his expertise in psychological assessments and consultations and Mr Brian Deakin of Capital Radiology Australia for technical assistance with the X-ray and CT imaging.

The suprachoroidal array and lead wire were designed by the Bionics Institute using research components supplied by Cochlear Ltd (Sydney, Australia).

AUTHOR CONTRIBUTIONS

Conceived and designed the experiments: LNA PJB RHG CDL DAXN NCS MNS MAP CDM JGW NB ANB CEW RKS PJA. Performed the experiments: LNA CDL DAXN NCS MNS JY MFM RJB NLO JV PND MV MAP CDM JGW NB PJA. Analyzed the data: LNA PJB RHG CDL DAXN NCS MNS JY MFM RJB NLO JV PND MV MAP CDM JGW NB ANB CEW RKS PJA. Contributed reagents/materials/analysis tools: PJB RHG NB ANB CEW RKS PJA. Wrote the paper: LNA PJB RHG CDL DAXN NCS MNS JY MFM RJB NLO JV PND MV MAP CDM JGW NB ANB CEW RKS PJA.

REFERENCES

1. Chader GJ, Weiland J, Humayun MS (2009) Artificial vision: needs, functioning, and testing of a retinal electronic prosthesis. Prog Brain Res 175:317–332.
2. Pascolini D, Mariotti SP (2012) Global estimates of visual impairment: 2010. Br J Ophthalmol 96:614–618.
3. Zrenner E (2013) Fighting blindness with microelectronics. Sci Transl Med 5:210ps216.
4. Shepherd RK, Shivdasani MN, Nayagam DA, Williams CE, Blamey PJ (2013) Visual prostheses for the blind. Trends Biotechnol 31:562–571.
5. Humayun MS, Dorn JD, Ahuja AK, Caspi A, Filley E, et al.. (2009) Preliminary 6 month results from the Argus II epiretinal prosthesis feasibility study. Conf Proc IEEE Eng Med Biol Soc: 4566–4568.
6. Rizzo JF 3rd (2011) Update on retinal prosthetic research: the Boston Retinal Implant Project. J Neuroophthalmol 31:160–168.
7. Zrenner E, Bartz-Schmidt KU, Benav H, Besch D, Bruckmann A, et al. (2011) Subretinal electronic chips allow blind patients to read letters and combine them to words. Proc Biol Sci 278:1489–1497.
8. Stingl K, Bartz-Schmidt KU, Besch D, Braun A, Bruckmann A, et al. (2013) Artificial vision with wirelessly powered subretinal electronic implant alpha-IMS. Proc Biol Sci 280:20130077.
9. Stone JL, Barlow WE, Humayun MS, de Juan E Jr, Milam AH (1992) Morphometric analysis of macular photoreceptors and

ganglion cells in retinas with retinitis pigmentosa. Arch Ophthalmol 110:1634–1639.

10. Bunker CH, Berson EL, Bromley WC, Hayes RP, Roderick TH (1984) Prevalence of retinitis pigmentosa in Maine. Am J Ophthalmol 97:357–365.
11. Santos A, Humayun MS, de Juan E Jr, Greenburg RJ, Marsh MJ, et al. (1997) Preservation of the inner retina in retinitis pigmentosa. A morphometric analysis. Arch Ophthalmol 115:511–515.
12. Hood DC, Lin CE, Lazow MA, Locke KG, Zhang X, et al. (2009) Thickness of receptor and post-receptor retinal layers in patients with retinitis pigmentosa measured with frequency-domain optical coherence tomography. Invest Ophthalmol Vis Sci 50:2328–2336.
13. Humayun MS, Prince M, de Juan E Jr, Barron Y, Moskowitz M, et al. (1999) Morphometric analysis of the extramacular retina from postmortem eyes with retinitis pigmentosa. Invest Ophthalmol Vis Sci 40:143–148.
14. da Cruz L, Coley BF, Dorn J, Merlini F, Filley E, et al.. (2013) The Argus II epiretinal prosthesis system allows letter and word reading and long-term function in patients with profound vision loss. Br J Ophthalmol: 632–636.
15. Humayun MS, Dorn JD, da Cruz L, Dagnelie G, Sahel JA, et al. (2012) Interim results from the international trial of Second Sight's visual prosthesis. Ophthalmology 119:779–788.
16. Fujikado T, Morimoto T, Kanda H, Kusaka S, Nakauchi K, et al. (2007) Evaluation of phosphenes elicited by extraocular stimulation in normals and by suprachoroidal-transretinal stimulation in patients with retinitis pigmentosa. Graefes Arch Clin Exp Ophthalmol 245:1411–1419.
17. Rizzo JF 3rd, Wyatt J, Loewenstein J, Kelly S, Shire D (2003) Methods and perceptual thresholds for short-term electrical stimulation of human retina with microelectrode arrays. Invest Ophthalmol Vis Sci 44:5355–5361.
18. Rizzo JF 3rd, Wyatt J, Loewenstein J, Kelly S, Shire D (2003) Perceptual efficacy of electrical stimulation of human retina with a microelectrode array during short-term surgical trials. Invest Ophthalmol Vis Sci 44:5362–5369.

19. Fujikado T, Kamei M, Sakaguchi H, Kanda H, Morimoto T, et al. (2011) Testing of semichronically implanted retinal prosthesis by suprachoroidal-transretinal stimulation in patients with retinitis pigmentosa. Invest Ophthalmol Vis Sci 52:4726–4733.
20. Second Sight Medical Products (2012) Sponsor Executive Summary: Prepared for the September 28, 2012 Meeting of the FDA Ophthalmic Device Advisory Panel. http://www.fda.gov/downloads/AdvisoryCommittees/CommitteesMeetingMaterials/MedicalDevices/MedicalDevicesAdvisoryCommittee/OphthalmicDevicesPanel/UCM320778.pdf: Accessed 11 December 2013.
21. Stingl K, Zrenner E (2013) Electronic approaches to restitute vision in patients with neurodegenerative diseases of the retina. Ophthalmic Res 50:215–220.
22. Ayton LN, Guymer RH, Luu CD (2013) Choroidal thickness profiles in retinitis pigmentosa. Clin Experiment Ophthalmol 41:396–403.
23. Villalobos J, Allen PJ, McCombe MF, Ulaganathan M, Zamir E, et al. (2012) Development of a surgical approach for a wide-view suprachoroidal retinal prosthesis: evaluation of implantation trauma. Graefes Arch Clin Exp Ophthalmol 250:399–407.
24. Villalobos J, Nayagam DA, Allen PJ, McKelvie P, Luu CD, et al. (2013) A wide-field suprachoroidal retinal prosthesis is stable and well tolerated following chronic implantation. Invest Ophthalmol Vis Sci 54:3751–3762.
25. Cicione R, Shivdasani MN, Fallon JB, Luu CD, Allen PJ, et al. (2012) Visual cortex responses to suprachoroidal electrical stimulation of the retina: effects of electrode return configuration. J Neural Eng 9:036009.
26. Shivdasani MN, Fallon JB, Luu CD, Cicione R, Allen PJ, et al. (2012) Visual cortex responses to single- and simultaneous multiple-electrode stimulation of the retina: implications for retinal prostheses. Invest Ophthalmol Vis Sci 53:6291–6300.
27. van den Honert C, Kelsall DC (2007) Focused intracochlear electric stimulation with phased array channels. J Acoust Soc Am 121:3703–3716.
28. Ayton LN, Apollo NV, Varsamidis M, Dimitrov PN, Guymer RH, et

al. (2014) Assessing residual visual function in severe vision loss. Invest Ophthalmol Vis Sci 55:1332–1338.

29. Finger RP, Tellis B, Crewe J, Keeffe JE, Ayton LN, et al. (2014) Developing the Impact of Vision Impairment-Very Low Vision (IVI-VLV) Questionnaire as Part of the LoVADA Protocol. Invest Ophthalmol Vis Sci 55:6150–6158.
30. Bittner AK, Haythornthwaite JA, Diener-West M, Dagnelie G (2013) Worse-than-usual visual fields measured in retinitis pigmentosa related to episodically decreased general health. Br J Ophthalmol 97:145–148.
31. Bittner AK, Ibrahim MA, Haythornthwaite JA, Diener-West M, Dagnelie G (2011) Vision test variability in retinitis pigmentosa and psychosocial factors. Optom Vis Sci 88:1496–1506.
32. Abramian M, Lovell NH, Morley JW, Suaning GJ, Dokos S (2011) Activation of retinal ganglion cells following epiretinal electrical stimulation with hexagonally arranged bipolar electrodes. J Neural Eng 8:035004.
33. Saunders AL, Williams CE, Heriot W, Briggs R, Yeoh J, et al. (2014) Development of a surgical procedure for implantation of a prototype suprachoroidal retinal prosthesis. Clin Experiment Ophthalmol 42:665–674.
34. John SE, Shivdasani MN, Leuenberger J, Fallon JB, Shepherd RK, et al. (2011) An automated system for rapid evaluation of high-density electrode arrays in neural prostheses. J Neural Eng 8:036011.
35. Bach M, Wilke M, Wilhelm B, Zrenner E, Wilke R (2010) Basic quantitative assessment of visual performance in patients with very low vision. Invest Ophthalmol Vis Sci 51:1255–1260.
36. Bach M (1996) The Freiburg Visual Acuity test—automatic measurement of visual acuity. Optom Vis Sci 73:49–53.
37. Dacey DM, Petersen MR (1992) Dendritic field size and morphology of midget and parasol ganglion cells of the human retina. Proc Natl Acad Sci U S A 89:9666–9670.
38. Lieby P, Barnes N, Walker JG, Scott AF, Barnes N, et al.. (2013) Evaluating Lanczos2 image filtering for visual acuity in simluated prosthetic vision. Association for Research in Vision and Ophthalmology (ARVO) annual meeting. Seattle, USA.

39. Barnes N, Scott AF, Stacey A, Lieby P, McCarthy C, et al.. (2014) Vision Processing with Lanczos2 Improves Low Vision Test Results in Implanted Visual Prosthetics. Association for Research in Vision and Ophthalmology annual meeting. Orlando, FL.
40. Nayagam DA, Williams RA, Allen PJ, Shivdasani MN, Luu CD, et al. (2014) Chronic electrical stimulation with a suprachoroidal retinal prosthesis: a preclinical safety and efficacy study. PLOS One 9:e97182.
41. Allen PJ, Yeoh J, McCombe MF, Heriot WJ, Luu CD, et al.. (2013) Implantation of a suprachoroidal retinal prosthesis results in number and letter recognition; Seattle, USA.
42. Huang CQ, Shepherd RK, Carter PM, Seligman PM, Tabor B (1999) Electrical stimulation of the auditory nerve: Direct current measurement in vivo. IEEE Trans Biomed Eng 46:461–470.
43. Merrill DR, Bikson M, Jefferys JG (2005) Electrical stimulation of excitable tissue: design of efficacious and safe protocols. J Neurosci Methods 141:171–198.
44. Blamey PJ, Sinclair NC, Slater K, McDermott HJ, Perera T, et al.. Psychophysics of a suprachoroidal retinal prosthesis; 2013; Seattle, USA.
45. Leung RT, Nayagam DA, Williams RA, Allen PJ, Salinas-La Rosa CM, et al.. (2014) Safety and efficacy of explanting or replacing suprachoroidal electrode arrays in a feline model. Clin Experiment Ophthalmol In press.
46. Bron A, Tripathi R, Tripathi B (1998) Wolff's Anatomy of the Eye and Orbit. London: Chapman & Hall.
47. Shivdasani MN, Sinclair NC, Dimitrov PN, Varsamidis M, Ayton LN, et al. (2014) Factors affecting perceptual thresholds in a suprachoroidal retinal prosthesis. Invest Ophthal Vis Sci 55:6467–6481.
48. Stingl K, Bartz-Schmidt KU, Gekeler F, Kusnyerik A, Sachs H, et al. (2013) Functional outcome in subretinal electronic implants depends on foveal eccentricity. Invest Ophthalmol Vis Sci 54:7658–7665.

CHAPTER 6

A SIMPLE EXOSKELETON THAT ASSISTS PLANTARFLEXION CAN REDUCE THE METABOLIC COST OF HUMAN WALKING

Philippe Malcolm, Wim Derave, Samuel Galle, and Dirk De Clercq

Department of Movement and Sports Sciences, Ghent University, Ghent, Belgium

ABSTRACT

Background

Even though walking can be sustained for great distances, considerable energy is required for plantarflexion around the instant of opposite leg heel contact. Different groups attempted to reduce metabolic cost with exoskeletons but none could achieve a reduction beyond the level of walking without exoskeleton, possibly because there is no consensus on the optimal actuation timing. The main research question of our study was whether it is possible to obtain a higher reduction in metabolic cost by tuning the actuation timing.

Methodology/Principal Findings

We measured metabolic cost by means of respiratory gas analysis. Test subjects walked with a simple pneumatic exoskeleton that assists plantarflexion with different actuation timings. We found that the exoskeleton can reduce metabolic cost by 0.18±0.06 W kg^{-1} or 6±2% (standard error of the mean) ($p=0.019$) below the cost of walking without exoskeleton if actuation starts just before opposite leg heel contact.

Conclusions/Significance

The optimum timing that we found concurs with the prediction from a mathematical model of walking. While the present exoskeleton was not ambulant, measurements of joint kinetics reveal that the required power could be recycled from knee extension deceleration work that occurs naturally during walking. This demonstrates that it is theoretically possible to build future ambulant exoskeletons that reduce metabolic cost, without power supply restrictions.

INTRODUCTION

Broader Context and Literature Review

At preferred speed, walking is the most metabolically economic gait mode [1] and can easily be sustained which allowed historic feats such as Roman legionnaires travelling 30 km per day [2]. This low metabolic cost is due to the total body center of mass (COM) moving as an inverted pendulum over the stance leg, thereby allowing interchange of potential and kinetic energy [3]. Still, walking has a substantial metabolic cost as there is only up to 70% energy conservation [3].

Different groups are developing assistive robotic devices, called exoskeletons, for carrying heavy loads, assisting the mobility of the world's aging population, *etc* [4]. Unfortunately, limited scientific information is available on the effects of these exoskeletons because of their commercial finality. In 2009, Ferris wrote a commentary article in which he expressed the need for basic studies on the physiology of exoskeleton assisted movement [5].

Several studies have been done with exoskeletons that assist plantarflexion [6]–[12]. This is a logical choice as around half of the positive muscle work during walking is delivered by the ankle [13]. However, none of these studies could produce reductions in metabolic cost below the level of normal walking without exoskeleton [6], [8], which implies that they have no practical benefit for sustained walking. Possible explanations could be that some exoskeletons are too heavy [14] and that there is no agreement on what would be the optimal actuation timing.

Aim and Hypothesis

The overall aim of our study was to find if it is possible to reduce metabolic cost below the level of normal walking without exoskeleton by changing actuation timing during the stance phase. The only phase of walking during which high positive joint work is required is at the end of the single stance phase, when the leading leg makes heel contact [13]. The ankle extensors of the trailing leg then serve to push the COM into the next inverted pendulum arc [15]. In addition, Kuo predicted by means of a simplified mathematical model that walking is most efficiently actuated with a push off just before opposite leg heel contact [16]. To achieve our aim we conducted controlled human experiments using bilateral exoskeletons that assist plantarflexion by means of pneumatic muscles [9].

Experimental Design

A computer program permitted to trigger the onset of actuation at five increments of the stride cycle (~13, 23, 34, 43 and 54%, (opposite leg heel contact occurs at 50%)) based on heel switch signals. Actuation offset was always at toe off (~63%). Eight subjects walked on a treadmill wearing the exoskeleton in the five onset conditions and a reference condition with unpowered exoskeleton. We calculated metabolic power based on respiratory gas analysis [17]. In order to account for differences in assistive power due to the different conditions we also expressed the metabolic effects as a ratio versus exoskeleton power. This ratio is called exoskeleton performance index [7].

RESULTS AND DISCUSSION

Metabolic Cost and Performance Index

We found a U-shaped pattern in metabolic cost versus actuation onset (Figure 1A). In the 43% onset condition we found the highest reduction of 0.64±0.05 W kg^{-1} (standard error of the mean (s.e.m.)) or 17±1% (s.e.m.) versus the net metabolic cost of the unpowered condition which is 3.72±0.19 W kg^{-1} (s.e.m.) ($p<0.001$, Tukey's honestly significant difference (THSD) versus unpowered condition). In the same condition we found the highest performance index ($p=0.006$, THSD versus 13% condition, Figure 1C). The observation that performance index is the highest in the condition with onset just before opposite leg heel contact concurs with the model of Kuo [16]. When we compared this optimal condition versus the net metabolic cost of walking without exoskeleton (3.25±0.11 W kg^{-1} (s.e.m.)) we found a reduction of 0.18±0.06 W kg^{-1} or 6±2% (s.e.m.) ($p=0.019$, paired t-test, Figure 1B).

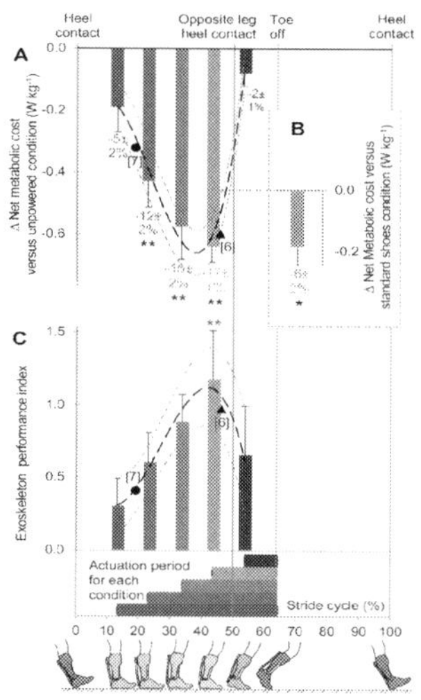

Figure 1. Metabolic cost and performance index. (A) Δ Net metabolic cost versus unpowered condition. Asterisks indicate significant differences versus unpowered condition. (B), 43% condition versus without exoskeleton. Asterisk indicates significant difference. (C) Performance index. Asterisks indicate significant difference versus 13% condition. Numbers below bars indicate differences expressed as percentages of net metabolic cost in unpowered or standard shoes condition. Horizontal bars indicate actuation duration. Vertical lines indicate heel contact and toe off. Filled circles (•) and triangles (▲) respectively indicate results derived [22], [35] from Sawicki and Ferris [7] and Norris et al. [6] (young adults population). Error bars indicate inter-subject s.e.m. Black and grey dashed lines indicate mean±s.e.m. of third-order polynomial curve fit. $^{**}p \leq 0.01$, $^{*}p \leq 0.05$.

Kinesiological Explanation

An explanation for the highest performance index in the 43% onset condition could be that the timing of the assistive power corresponds best to the biological positive ankle power as described in the literature [13], or as observed in inverse dynamical analyzes with our exoskeleton. It is remarkable that the optimal onset (at 43%) occurred much later than the onset of biological plantarflexor activation which occurs around 15% of the stride cycle [18]. This is probably because biological plantarflexors produce negative work by lengthening during the first part of the stance phase [18]. The exoskeleton produced almost exclusively positive work even in the earliest onset conditions by forcing the ankle angular velocity into plantarflexion.

As such, our results suggest that a steering method based purely on biological muscle activation is not ideal for reducing metabolic cost during steady state walking with plantarflexion assisting exoskeletons with concentric actuation. The fact that Norris et al. [6] found a higher reduction in metabolic cost than Sawicki and Ferris [7] despite a much shorter habituation period (respectively 2×5 min instead of 3×30 min) could be interpreted from this perspective.

Explanation by Step-to-Step Transition

In order to investigate the link with the inverted pendulum model we looked at the kinematics.We found that in the condition with the highest reduction in metabolic cost (*i.e.* 43% onset condition), the drop of the COM during the double stance phase (2.7±0.7 mm (s.e.m.)) was around half the size of the COM drop in 13% onset condition (5.9±0.9 mm (s.e.m.), p=0.010, THSD versus 13% condition, Figure 2B). We also found a significant correlation of net metabolic cost with COM drop during the double stance phase (p=0.015, R=−0.38, Pearson's correlation of Δ's versus unpowered condition). There was no significant correlation with vertical COM excursion during the whole stride (p=0.832, R=0.04).

As such, the lower metabolic cost in the optimal onset condition can be attributed specifically to the COM being more effectively redirected during the step-to-step transition which experimentally validates the model of Kuo [16]. Actuation onset had no effect on the spatiotemporal parameters stride length and stride time (p>0.593, repeated measures analysis of variance) which could otherwise have confounded the effect of COM drop on the step-to-step transition.

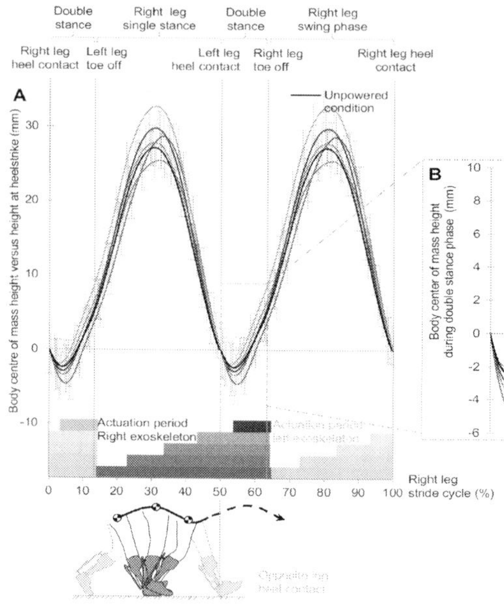

Figure 2. Body center of mass (COM) height. (A) COM height versus height at heelstrike during right leg stride cycle. Error bars indicate inter-subject s.e.m. Opaque horizontal bars indicate actuation duration of right leg exoskeleton. Transparent bars indicate the actuation duration of opposite leg exoskeleton. Vertical lines delimit single & double stance phases. (B) COM height during double stance phase. Arrows indicate COM drop after heel contact. It can be noted that the arrows are slightly larger than the minima of the lines in the chart. This due to the temporal variation of the occurrence of minimum COM height is which is not shown in the lines in the chart as these only show the mean evolution of COM height. Asterisk indicates significant Pearson's correlation between Δ COM drop versus unpowered condition and Δ net metabolic cost versus unpowered condition. Black line and arrow indicate unpowered condition. *p≤0.05.

Actuation Timing Guidelines

From a practical perspective our results could be used as guidelines for steering exoskeletons. Curve fitting of the results of the tested conditions shows the highest reduction in metabolic cost could take place with onset at 37±1% (s.e.m.). This timing could be useful when power availability is not an issue, such as during rehabilitation exercise on treadmill. Also by using curve fitting, the highest performance index could be expected with onset at 45±2% (s.e.m.). This timing could be used for ambulant exoskeletons for which efficient use of power is critical. While no other study compared multiple actuation timings within the same subjects, multiple references comply with the relationship between actuation timing and metabolic parameters that we found. Results from Sawicki and Ferris [7] and Norris et al. [6] closely fit to the patterns of metabolic cost and performance index in our study (Figure 1A, C). Patients with incomplete spinal chord injuries walking with a plantarflexion assisting orthosis which they control with pushbuttons spontaneously adopt a timing that results in onset of plantarflexor torque at 43.5±3.7% (s.d.) of the stride cycle [10]. Transtibial amputees walking with a powered prosthesis subjectively indicate that opposite leg heel contact is the best time for adding power [19]. A simulation of an elastic exoskeleton shows that actuator stiffness is optimal when it allows plantarflexion to start just prior to opposite leg heel contact [20]. Altogether this suggests that an onset of concentric actuation somewhere between 40 to 50% is the most efficient for different plantarflexion assisting devices in different populations.

Practical Significance

To illustrate the significance of the reduction that we found in net metabolic cost versus walking without exoskeleton of 6±2% (s.e.m.) this could be compared to the effect on metabolic cost of supporting 25±8% (s.e.m.) of the body weight [21]. The ability to obtain such a reduction merely by briefly acting on a distal joint could be useful to augment endurance in able bodied subjects (*e.g.* rescue workers) or to restore performance in impaired subjects (*e.g.* elderly) [4]. Based on regression formulas from the literature [22], the present reduction would allow an increase in speed of 0.05±0.02 m s^{-1} (s.e.m.), which can be categorized as a small meaningful improvement in elderly [23]. By trimming the weight of the exoskeleton (0.67 kg per side) to a realistic weight of commercial ankle-foot orthoses (0.40 kg [24]) a further reduction in metabolic cost of ~3% [25] could be possible.

Hypotheses on Feasibility of Ambulant Exoskeletons

The most important limitation of our exoskeleton is that it is tethered to a non-portable compressed air supply. It will probably take some time until ambulant exoskeletons will reach the same level of performance in able bodied subjects.Current studies with ambulant exoskeletons report no reductions but increases in metabolic cost going from 10 to 60% [26]–[28]. The main obstacle is power source portability [4]. It has been suggested that recycling negative work, similar to regenerative braking in hybrid cars, could be a solution for this. This solution has been proposed for the design of powered prostheses [29] and exoskeletons [12], [20], [28], [30]. While in the design of the recycling prosthesis the authors had the liberty to add moving parts wherever they wanted, in the design of a recycling exoskeleton one can only use joints that map to the able bodied human anatomy. Another limitation for exoskeletons is that only work that would otherwise be dissipated as heat should be recycled. Negative work phases followed by positive work should not be used as these are used for biological storage and return of elastic tendon energy [31]. Surprisingly, current designs for recycling exoskeletons [12], [20], [26], [28], [30] do not yet take this into account.

In order to evaluate if sufficient negative joint work would be available, we did a supplementary inverse dynamic analysis in eight subjects with the exoskeleton operating in the 43% onset condition. In the knee power we found a marked negative peak near the end of the swing phase that is not used for storage and return of tendon energy as it is followed by another negative peak [31] (Figure 3). The absolute value of the work during this phase (0.24±0.01 J kg^{-1}(s.e.m.)) was found to be significantly higher (p=0.001, two-sample t-test) than the positive exoskeleton work in the 43% onset condition (0.11±0.03 J kg^{-1} (s.e.m.)) which means that sufficient work is available.

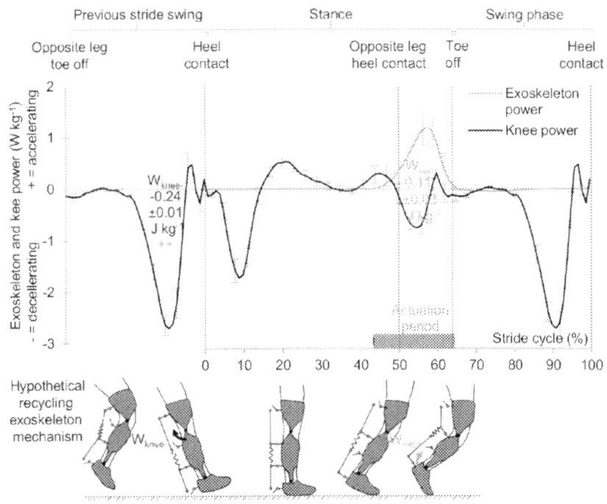

Figure 3. Hypothetical feasibility of an ambulant recycling exoskeleton. Black and green line respectively show knee and exoskeleton power in 43% onset condition. Numbers in chart surface indicate work during certain power peaks. Asterisks indicate significant difference between absolute value of knee swing deceleration work (W_{knee-}) versus positive exoskeleton work (W_{exo+}). Knee power is shown starting from the previous stride swing phase to illustrate how knee swing deceleration could be recycled into plantarflexion power during push off with a mechanism. Horizontal bar indicates actuation duration. Vertical lines indicate heel contact and toe off. Error bars indicate inter-subject s.e.m. **$p \leq 0.01$.

Future research should be oriented towards modelling of a lightweight mechanism (*e.g.* Figure 3) that can capture energy from knee swing deceleration, transfer it to the ankle and release it with optimal timing and minimal energy loss. Finally, prototype tests should be done with subjects as the human response to walking with an exoskeleton remains unpredictable [11] and the gains could be offset by the additional encumbrance. On the other hand, the substitution of negative work by the exoskeleton could provide small metabolic gains and users could also obtain additional gains by adapting their gait to optimally exploit the exoskeleton [28], [30], just like they are able to exploit biological features like biarticular muscles [18]. This training period will probably take more time than with our pneumatic exoskeleton as a recent study with an elastic exoskeleton shows continuous improvements in metabolic cost after each day of a four day training period [12].

Conclusion

The combination of the finding that it is possible to reduce metabolic cost with a pneumatic exoskeleton and the finding that sufficient naturally occurring negative joint work is available to reproduce the power of the pneumatic exoskeleton demonstrates that it is theoretically possible to build ambulant exoskeletons that improve the metabolic economy of walking without power supply restrictions. Once this is achieved, exoskeletons could become practically useful and start to appear in everyday life as predicted by Ferris [5] and Herr [14].

MATERIALS AND METHODS

Ethics Statement

The experiment was approved by the ethics committee of the Ghent University hospital (Belgian registration number B670220097074) and written informed consent was obtained from all subjects.

Exoskeletons

The subjects were equipped with bilateral hinged ankle foot exoskeletons [9] that were worn with regular shoes and powered by McKibben-type pneumatic muscles (28 cm contractile length, 3 cm diameter in relaxed state). A computer program (Labview, National Instruments, Austin, TX, USA) permitted to trigger the onset and offset of the pneumatic muscle contraction at predetermined percentages of the stride cycle based on an algorithm that predicted stride time from heel switches (IP67, Herga Electric, Suffolk, UK). The inflation pressure was 3.5 bar.

Conditions

The onset of the actuation was set at different increments of the stride cycle (~13, 23, 34, 43 and 54%) in five conditions. The offset of the actuation was fixed at toe off (~63%). In the unpowered condition the subjects walked with the exoskeleton without pneumatic muscle actuation. In the standard shoes condition the subjects walked without exoskeleton but with running shoes as in Norris et al. [6]. The different onset conditions and the unpowered condition were randomized. The standard shoe condition was semi-randomized, *i.e.* it was done alternatively before or after the exoskeleton conditions.

Subjects

We tested 10 subjects (♀, 23±1 years, 1.70±0.03 m, 66±4 kg (s.e.m.)) during treadmill walking at 1.38 m s^{-1}. In a subsample of 8 subjects we recorded metabolic, kinematic, spatiotemporal and exoskeleton parameters in the five actuation onset conditions and the unpowered condition. In another subsample of 8 subjects we recorded metabolic cost during walking with standard shoes without exoskeleton and during walking with the exoskeleton operating in the 43% onset condition. In a separate sample of 8 subjects (7♀/1♂, 21±0 years, 1.67±0.02 m, 60±1 kg (s.e.m.)) we performed inverse dynamic analyzes with the exoskeleton operating in the 43% onset and unpowered condition.

Habituation

Before the measurements subjects received at least five minutes habituation to walking on treadmill and another five minutes to walking with the exoskeleton. This could be a limitation of our protocol as other studies used at least two times five minutes [6] of habituation to walking with exoskeleton. However, as mentioned in the discussion on the kinesiological explanation, exoskeletons with sufficiently sophisticated spatiotemporal control could require less habituation than exoskeletons with EMG control [32] so our subjects were probably reasonably well habituated.

Joint Kinematics

In the treadmill experiments we recorded saggital kinematics of the right ankle joint, hip and pneumatic muscle endpoints by tracking (Maxtraq, Innovision Systems, Columbiaville, MI, USA) reflective markers from images of a 60 Hz camera (Basler AG, Ahrensburg, Germany). In the overground experiments we recorded full body 3D kinematics with a motion capture system at a rate of 200 Hz (Qualisys, Gothenburg, Sweden). We filtered the kinematics with a fourth order Butterworth lowpass filter with a 12 Hz cutoff frequency.

Exoskeleton Kinetics

We measured the tensile force of the pneumatic muscles at a rate of 1000 Hz by means of a load cell (W2, A.L. Design, Buffalo, NY, USA) and filtered it with a fourth order Butterworth lowpass filter with a 12 Hz cutoff frequency. Actuation onset was defined as the instant when the pneumatic valve was open and pneumatic muscle force exceeded a threshold of 5 N. All the results in this paper show the measured onset so they are not affected by a potential latency of the steering system. We calculated exoskeleton power by multiplying ankle angular velocity with the pneumatic muscle moment. The pneumatic muscle moment was obtained by multiplying the tensile force with the moment arm versus the ankle.

Metabolic Cost and Exoskeleton Performance Index

We analyzed respiratory gasses from the last two minutes of each four minute treadmill condition with a computerized O_2-CO_2 analyzer-flow meter (Oxycon Pro, Jaeger GMBH, Höchberg, Germany). We estimated metabolic cost with the formula from Brockway [17]. We calculated net metabolic cost by subtracting the metabolic cost measured during the last two minutes of four minutes standing still before the experiments from the metabolic cost in the experimental conditions. In order to specifically reflect the effects of the assistance from the exoskeleton, results were reported as the difference versus the unpowered condition and versus the standard shoes condition. In order to reflect the efficiency of the exoskeleton we calculated an index proposed by Sawicki and Ferris [7].

$$\text{Performance index} = \frac{0.25 \times \Delta\text{Net metabolic cost versus unpowered condition}}{\text{Mean positive power from bilateral exoskeletons}} \quad (1)$$

Body Center of Mass Height

We estimated the change in COM height based on the change in vertical position of pelvis [33]. The change in vertical position of the pelvis was calculated by taking the mean of the vertical position of the right hip and the estimated vertical position of the left hip obtained from shifting the right hip vertical position by half a stride.

Joint Kinetics

In order to be able to measure ground reaction forces we let subjects walk over at walkway with an embedded forceplate (Kistler, Winterthur, Switzerland). A similar walking speed as in the treadmill experiments was imposed by means of pacing lights alongside the walkway. The compressed air supply and the exoskeleton steering unit were rolled in a cart behind the subject. Sagittal ankle and knee joint kinetics were calculated with dedicated software (Visual3D, C-motion, Rockville, MD, USA) based on ground reaction forces, 3D kinematics and a segmental mass distribution model from Dempster [34] that was modified to incorporate the parts of the exoskeleton.

Spatiotemporal Parameters

In the treadmill experiment we determined initial contact and toe off from images of the 60 Hz camera. Step length was derived based on the treadmill speed. In the overground experiments we determined initial contact times based on the ground reaction forces and kinematics.

Statistics

We opted for parametric statistics in order to obtain comparable results as other studies on metabolic cost of walking with plantarflexion assisting exoskeletons with similar sample sizes (*i.e.* 9 subjects) [6]–[8]. In the first subsample of subjects we analyzed the overall effect of actuation timing on net metabolic cost, performance index and COM drop with a repeated measures analysis of variance. In order to estimate at what onset timing the highest reduction in metabolic cost and the highest performance index would be situated between the tested conditions we calculated the maxima of third order polynomial curve fits versus actuation timing (R^2=0. 91±0.06 (s.e.m.) for metabolic cost and R^2=0.80±0.07 (s.e.m.) for performance index, coefficients of determination). Differences between conditions were analyzed with a Tukey's honestly significant difference test. Correlations of metabolic cost versus COM kinematic parameters were analyzed with Pearson's correlation. In the second subsample of subjects we analyzed the difference in metabolic cost of the 43% onset condition versus the standard shoes condition with a paired t-test. In the sample of subjects who were tested overground we compared absolute value of the knee swing deceleration negative work versus the exoskeleton positive work from the optimal 43% onset condition from the treadmill experiments with a two-sample t-test. For all tests we used 8 subjects, two-tailed p-values and an alpha level of 0.05.

ACKNOWLEDGMENTS

The authors wish to thank Technische Orthopedie België for their help with constructing the exoskeleton and Davy Spiessens, Maarten Afschrift, Robin Decottignies and Sander De Bruyne for their help with the study.

AUTHOR CONTRIBUTIONS

Conceived and designed the experiments: PM WD SG DDC. Performed the experiments: PM SG. Analyzed the data: PM WD SG DDC. Contributed reagents/materials/analysis tools: PM WD SG. Wrote the paper: PM WD SG DDC.

REFERENCES

1. Hreljac A (1993) Preferred and energetically optimal gait transition speeds in human locomotion. Med Sci Sports Exerc 25: 1158–1162.
2. Fornaris E, Aubert M (1998) Le légionnaire romain, cet athlète méconnu. Hist Sci Med 32: 161–168.
3. Cavagna GA, Heglund NC, Taylor CR (1977) Mechanical work in terrestrial locomotion: two basic mechanisms for minimizing energy expenditure. Am J Physiol 233: R243–R261.
4. Bogue R (2009) Exoskeletons and robotic prosthetics: a review of recent developments. Ind Rob 36: 421–427 doi http://dx. doi. org/10. 1108/01439910910980141.
5. Ferris DP (2009) The exoskeletons are here. J Neuroeng Rehabil 6: 17 .
6. Norris JA, Granata KP, Mitros MR, Byrne EM, Marsh AP (2007) Effect of augmented plantarflexion power on preferred walking speed and economy in young and older adults. Gait Posture 25: 620–627 .

7. Sawicki GS, Ferris DP (2008) Mechanics and energetics of level walking with powered ankle exoskeletons. J Exp Biol 211: 1402–1413 .
8. Sawicki GS, Ferris DP (2009) Powered ankle exoskeletons reveal the metabolic cost of plantar flexor mechanical work during walking with longer steps at constant step frequency. J Exp Biol 212: 21–31 .
9. Malcolm P, Fiers P, Segers V, Van Caekenberghe I, Lenoir M, et al. (2009) Experimental study on the role of the ankle push off in the walk-to-run transition by means of a powered ankle-foot-exoskeleton. Gait Posture 30: 322–327 .
10. Sawicki GS, Domingo A, Ferris DP (2006) The effects of powered ankle-foot orthoses on joint kinematics and muscle activation during walking in individuals with incomplete spinal cord injury. J Neuroeng Rehabil 3: 3 .
11. Gordon KE, Sawicki GS, Ferris DP (2006) Mechanical performance of artificial pneumatic muscles to power an ankle-foot orthosis. J Biomech 39: 1832–1841 .
12. Wiggin MB, Collins SH, Sawicki GS (2012) A Passive Elastic Ankle Exoskeleton Using Controlled Energy Storage and Release to Reduce the Metabolic Cost of Walking. Proceedings 7th Annual Dynamic Walking Conference. 24–25.
13. Winter DA (1983) Energy generation and absorption at the ankle and knee during fast, natural, and slow cadences. Clin Orthop Relat Res: 147–154.
14. Herr H (2009) Exoskeletons and orthoses: classification, design challenges and future directions. J Neuroeng Rehabil 6: 21 .
15. Ruina A, Bertram JEA, Srinivasan M (2005) A collisional model of the energetic cost of support work qualitatively explains leg sequencing in walking and galloping, pseudo-elastic leg behavior in running and the walk-to-run transition. J Theor Biol 237: 170–192 .
16. Kuo AD (2002) Energetics of actively powered locomotion using the simplest walking model. J Biomech Eng 124: 113–120 .
17. Brockway JM (1987) Derivation of formulae used to calculate energy expenditure in man. Hum Nutr Clin Nutr 41: 463–471.

18. Ishikawa M, Komi P V, Grey MJ, Lepola V, Bruggemann GP (2005) Muscle-tendon interaction and elastic energy usage in human walking. J Appl Physiol 99: 603–608 .
19. Au SK, Dilworth P, Herr H (2006) An ankle-foot emulation system for the study of human walking biomechanics. Proceedings of the IEEE International Conference on Robotics and Automation. 2939–2945. doi:http://dx. doi. org/10. 1109/ROBOT. 2006. 1642148.
20. Bregman DJJ, Van der Krogt MM, De Groot V, Harlaar J, Wisse M, et al. (2011) The effect of ankle foot orthosis stiffness on the energy cost of walking: A simulation study. Clin Biomech 26: 955–961 doi http://dx. doi. org/10. 1016/j. clinbiomech. 2011. 05. 007.
21. Grabowski A, Farley CT, Kram R (2005) Independent metabolic costs of supporting body weight and accelerating body mass during walking. J Appl Physiol 98: 579–583 .
22. Givoni B, Goldman RF (1971) Predicting metabolic energy cost. J Appl Physiol 30: 429–433.
23. Perera S, Mody SH, Woodman RC (2006) Meaningful change and responsiveness in common physical performance measures in older adults. J Am Geriatr Soc 54: 743–749 .
24. Smith AE, Quigley M, Waters R (1982) Kinematic Comparison of the BiCaal Orthosis and the Rigid Polypropylene Orthosis in Stroke Patients. Orthot Prosthet 36: 49–55.
25. Browning RC, Modica JR, Kram R (2007) The effects of adding mass to the legs on the energetics and biomechanics of walking. Med Sci Sports Exerc 39: 515–525 .
26. Walsh CJ, Endo K, Herr H (2007) A quasi-passive leg exoskeleton for load-carrying augmentation. Int J HR 4: 487–506.
27. Gregorczyk KN, Hasselquist L, Schiffman JM, Bensel CK, Obusek JP, et al. (2010) Effects of a lower-body exoskeleton device on metabolic cost and gait biomechanics during load carriage. Ergonomics 53: 1263–1275 .
28. Van Dijk W, Van der Kooij H, Hekman E (2011) A Passive Exoskeleton with Artificial Tendons. Proceedings of the IEEE International Conference on Rehabilitation Robotics. 1–6. doi:http://dx. doi. org/10. 1109/ICORR. 2011. 5975470.

29. Collins SH, Kuo AD (2010) Recycling energy to restore impaired ankle function during human walking. PLoS one 5: e9307 .
30. Van den Bogert AJ (2003) Exotendons for assistance of human locomotion. Biomed Eng Online 2: 17 .
31. Qingguo L, Naing V, Maxwell DJ (2009) Development of a biomechanical energy harvester. J Neuroeng Rehabil 6: 22 .
32. Galle S, Malcolm P, Derave W, De Clercq D (2013) Adaptation to walking with an exoskeleton that assists ankle extension. Gait Posture.
33. Thirunarayan MA, Kerrigan DC, Rabuffetti M, Croce UD, Saini M (1996) Comparison of three methods for estimating vertical displacement of center of mass during level walking in patients. Gait Posture 4: 306–314 doi http://dx. doi. org/10. 1016/0966-6362(95)01058-0.
34. Dempster WT (1955) Space requirements of the seated operator. WADC Tech Rep: 55–159.
35. Mifflin MD, St Jeor ST, Hill LA, Scott BJ, Daugherty SA, et al. (1990) A new predictive equation for resting energy expenditure in healthy individuals. Am J Clin Nutr 51: 241–247.

CHAPTER 7

PLANT BIONICS

Prerna Pandey[1] and Shiv Sanjeevi Sripathi[2]

[1]Dr Prerna Pandey, a biotechnologist with several years of wet lab research experience, worked at International Center for Genetic Engineering and Biotechnology, New Delhi. Her field of research in PhD included isolation and molecular characterization of geminiviruses, genome sequencing, gene annotation, and gene silencing using the RNA interference technology. She has also worked at Transasia Biomedicals and Advance Enzyme Technologies as a scientist. Prerna has published papers in peer-reviewed journals, and has submitted a number of annotated Geminiviral genome sequences in the GenBank, including two novel ones. She has also completed her editing and proof-reading courses from Society for the promotion of Editors and Proof-readers, UK and now works as a freelance scientific editor and writer. When Prerna is not busy with her assignments, she loves to explore historical places.

[1]Shiv Sanjeevi Sripathi completed his BSc in Biotechnology from Mumbai University and secured second rank in Mumbai University and first rank in his college: Kishinchand chellaram College. Shiv secured first rank in his college KET's V.G.Vaze College and has worked on a stem cell project at the Specialized Centre for Cell Based Therapy (SCCT), KEM Hospital at Mumbai, following which he was a Junior Research Fellow at Microbiology & Cell Biology Department, Indian Institute of Science, Bangalore from August 2009 to November 2011. The field of work involved cloning of cell wall genes and transcription factors in E. coli & M. smegmatis. This was followed by a stint as a Research and Development Executive at Transasia Biomedicals, Mumbai where he worked on patent writing and developing immunoassays from 2011 till 2013. In 2013,

he entered the teaching field where he has been tutoring biology and chemistry to the students of 11th and 12th grades along with English grammar and communication skills to people of all age groups.

INTRODUCTION

The most heated research subject Bionics is widely used in many fields, from military, medicine, industry to building, information and so on. It is one of the interdisciplinary areas, combining the life sciences with the engineering sciences, in other words, the art of applying the knowledge of the functions of living systems to solving technical problems. Bionics coined by E. Steele of the Aerospace Division of the U.S. Air Force, from the Greek word for life and suffix "ic," meaning "having the nature of" in August 1958. These days, major progress in material science is mae possible by taking inspiration from nature. The models of the reference biological systems provide cues to the design of different components of an artificial system.

Plants or plant parts offer countless clues for making innovation in technology. Plants have evolved the capability to respond to a wide range of signals and are able to show considerable plasticity in their morphology and physiology in response to variability within their environment. Plant materials are optimized to reduce energy consumption during motion and these capabilities offer a plethora of solutions in the artificial world, although plants lack muscles, they generate several different kinds of movements that span from hours or days to milliseconds (Martone et al., 2010; Forterre, 2013). An understanding of these non-muscular movements could hold the potential for developments in applied sciences and engineering, especially for the fabrication of novel biomimetic actuation strategies and bioinspired devices (Burgert and Fratzl, 2009).

A few examples are described to lay the perspectives of plants in the artificial world. The invention of Velcro is one of the most famous example of technology inspired by plants, when in 1948, a Swiss engineer, George de Mestral, noticed the hooks of the plant burrs (*Arctium lappa*) stuck in the fur of his dog. From this observation, George de Mestral derived the idea of a novel type of zip fastener (Velcro, 1955). Another interesting industrial solution resulting from the study of material properties in plants is Lotusan®, a paint for selfcleaning surfaces. The scientific basis of this invention is the so-called lotus effect: (Barthlott and Neinhuis, 1997; Bhushan, 2009). Concerning intrinsic mechanical structural properties, an interesting example of technological solution inspired by plants is given by cellular or sandwich structures consisting of solid shells filled with compliant cellular cores, which have increased resistance against axial buckling while being light weight. These properties are typically required in aerospace applications (Meyers et al., 2008). Another example is given by the pomello, *Citrus maxima*, which shows a peculiar damping system, allowing the fruit to resist to a drop from heights of up to 15 m. The complex wall hierarchical structure consisting of interconnected porous layers combined with branched fiber networks, is able to dissipate more than 90% of the impact energy thus preventing pulp damage (Martone et al., 2010; Thielen et al., 2013). The principles have been exploited in the development of new materials for enhanced crash absorbers, based on metal foams (Fischer et al., 2010).

MECHANICAL CHARACTERISTICS OF TYPICAL PLANT LEAVES

The microstructure is a major contributing factor in the strength of a plant leaf, the natural composite biomaterial. Wang et al (2010) analysed the mechanical characteristics of eight species of plant leaves and measured the ultimate strength and the hardness of leaf surfaces by using universal testing machine and nanoindenter tester, respectively. The results indicate that the material composition and structure define the ultimate strength of a leaf. The results of hardness tests revealed that the surface hardness is influenced by the components of leaf surface. A higher surface hardness is shown by leaf surfaces composed of more lignin and cellulose than that composed of more carbohydrates.

PLANT NANOBIONICS

In the field of agriculture, the use of nanomaterials (NMs) would improve by making the agricultural practices more efficient and sustainable by generating less waste than conventional practices. In this perspective, modern nanotechnology in agriculture is considered as one of the important approaches to significantly enhance crop productivity. Nutrients that are nano-encapsulated, or, nanofertilizers can effectively to release nutrients on demand to regulate plant growth and enhance productivity (Rosa et al., 2010; Nair et al., 2010). While the use of NMs is comparatively new in the field of agriculture, it needs further exploration.

Plants for the Biological Synthesis of Nanoparticles

The biological methods of nanoparticle synthesis follow "Bottom Up" approaches in which the nanoparticles are built from small entities by reduction and oxidation processes and thus are obtained with lesser defects (Singha et al., 2015). The biological entities possess capping and stabilizing agents in the synthesis process. Bio-reduction of NMs takes place either in-vitro or in-vivo. During synthesis, enzymes, proteins, sugars, and the cofactors, flavonoids, terpenoids, phenolics, etc. mainly act as reducing and stabilizing agents.

- The leaf extract of *Eclipta prostrata* was used for the Synthesis of TiO2 NPs (36-38 nm, spherical) from titanium hydroxide TiO (OH)2 solution at room temperature.
- CuO nanoparticles (5-10 nm, spherical) were found to be active against both gram positive and gram-negative bacteria and could be synthesised by reduction using *Malva sylvestris*. In addition, when *Phyllanthus amarus* leaf extract was exposed to copper sulfate (Cu2SO4) solution, with vigorous stirring for 7 hours at 130 °C, CuO nanoparticles (20 nm, spherical) could be synthesized, which exhibited antibacterial activity against *Bacillus subtilis* compared to rifampicin.
- *Gloriosa superba* leaf extract was used to synthesize CeO2 NPs (5 nm, spherical) with antibacterial properties.
- The leaf extracts of green tea and eucalyptus produced Fe nanoparticles (20-80 nm, spherical) which showed significant waste water in situ remediation, especially in nitrate contaminated sites.
- The extracellular synthesis of Ag, Au and Au-Ag NPs (20-150 nm, triangular, spherical, and hexagonal) in water was reported by, using an edible mushroom *Volvariella volvacea* extract.

The size, kinds, chemical composition and stability of the NPs are defined by the uptake, translocation, and accumulation of these NPs in the plant species (Reviewed by Singha et al., 2015).

Introduction of Artificial Electroactive Materials in Plants

Now a day analog and digital organic electronic circuits and devices manufactured in living plants. The roots, stems, leaves, and vascular circuitry of higher plants are analogous to the contacts, interconnections, devices, and wires of discrete and integrated electronic circuits.

Introduction of artificial electroactive materials in living plants have been reported. A very apt example of living plants hosting analog and digital organic electronic circuits and devices was shown in *Rosa floribunda* (garden rose) (Stavrinidou et al., 2015). The electronically conducting polymer poly(3,4-ethylenedioxythiophene) (PEDOT), is an organic electronic material that typically exhibit high combined electronic and ionic conductivity in the hydrated state. *R. floribunda* stem was cut, and the fresh cross section was immersed in an aqueous PEDOT-S:H solution for 24 to 48 hours (for the solution to be taken up into the xylem vascular channel and transported apically), the result was a leaf composed of a two-dimensional (2D) network of compartments – either fully or partly filled - with the electronicionic material. The analysis showed homogeneous electrochromism in the compartments in direct stomatal contact with the external PEDOT:PSS–NFC electrodes (Stavrinidou et al., 2015). This indicates that ion conduction across veins is efficient and does not limit the overall charge transport.

The xylem, veins, leaves and signals of plants help achieve the components of a circuit that are important and act as integral part of the circuit elements and functions. Thus, ion migration pathways within a leaf can be efficiently studied using induced electrochromism and optical image analysis.

The Growth Promoting Nanoparticles

The nanoparticles have affect the plant growth and development both positively and negatively. There have been several reports of the effects of nanomaterials on germination and growth with the goal to promote their use in agriculture. The nanoparticle efficiency is determined by their surface area, chemical composition, reactivity, size, and the concentration at which they response positively.

When barley, soybean and corn were treated with Multi-walled carbon nanotubes (MWCNTs), seed germination increased (Lahiani et al., 2013). Oxidized MWCNTs increased cell elongation in root system and promoted dehydrogenase activity (Wang et al., 2012). The exposure of single-walled carbon nanotubes (SWCNTs) was reported by Canas et al (2008) to enhance the growth of root of onion and cucumber seedlings.

Several positive changes were observed when the plants were exposed to specific nanoparticles that were biologically synthesized including seed germination in *Boswellia avalifoliata;* increased synthesis of protein and carbohydrate and decreased the total phenol content *in Baopa monnieri*; increased root length in barley; more than two-fold increase in height and fresh weight in duckweed; significantly enhanced shoot, chlorophyll content and grain yield in pearl millet (Reviewed by Singha et al., 2015). In hydroponic solution, Almeelbi and bezbaruah (2014) reported significant enhancement in plant growth and biomass on using Fe nanoparticles on spinach.

A large number of studies on the effects of metal oxide NPs on **germination and growth** of plants have been documented in parsley seeds, salvia seeds, soybean, spinach, pumpkin, peanut plant, maize seeds, Changbai larch (Larix olgensis) seedlings, seed germination in wheat, *Cyamopsis tetragonoloba* in several growth parameters, canola seedlings (Reviewed by Singha et al., 2015).

In wheat, chlorophyll formation and activities of Rubisco was promoted by TiO2 nanoparticles (Yang et al., 2006). Root elongation and photosynthetic potential was seen to be significantly enhanced by application of nanoparticles on the surface, while far less enhancement was seen when NPs were given to plant via soil route; a major cause of this could be the precipitation of Fe ions.

Bionic Plants as Chemical Monitors of Groundwater Contamination

The living, wild-type plants can function as chemical monitors of groundwater and communication devices to external electronics. Especially designed nanoparticles can be embedded into plants to impart them with specific, non-native functions. Researchers at the Massachusetts Institute of Technology in Cambridge coated carbon nanotubes with a peptide that binds to nitroaromatic compounds, also contained in the explosives (Wong et al., 2017). Live spinach plants (*Spinacia oleracea*) can be engineered to serve as platforms that send infrared signals to a smartphone. These Nano-bionic plants are self-powered, capable of autosampling analytes by employing a pair of near-infrared fluorescent nanosensors. These leaf mesophyll-embedded sensors are peptide Bombolitin II conjugated single-walled carbon nanotubes (SWCNTs) that recognize nitroaromatics via infrared fluorescent emission, and the invariant reference signal is given by polyvinyl-alcohol functionalized SWCNTs. The fluorescence emission changes in the nanotubes when chemical contaminants attach to the nanotubes as they are absorbed by the roots or leaves. This decrease is proportional to the amount of the compound. The signal is picked up by a small detector which sends it wirelessly to a smartphone. The real-time monitoring of embedded SWCNT sensors also allows estimation of residence times of the contaminants in the roots, stems and leaves.

The bionic blades can provide guidance for designing cutting component of succulence cutter, vegetable choppers, and other machines used in food processing. Plant fiber can be very efficiently cut with bamboo weevil larva. The cutting efficiency of mandible incisor is greatly improved because the primary cutting edge of larval incisor is close to a standard circular arc. Tong et al (2017) have reported the design and manufacture of a new bionic mincing blade that is inspired from the weevil larva's incisor characteristics. Wong et al showed through chopping equal Chinese cabbages that, at 1400 rpm rotational speed, the mincing energy consumption of bionic blade was 12.8% lower than conventional blade and the chopping efficiency of bionic blade was 12.5% higher. Compared to a conventional blade, the material weight loss rate was 11.2 % lower and the qualification rate of the minced cabbage chopped by bionic blade was 93.3%, higher than that of conventional blade at 85.7%. The bionic blade thus promoted product quality and efficiency of mincing machine.

Although several plant leaf spectra simulation models have been proposed, material simulation of plant leaves has been rarely reported. The reflectance spectra of plant leaves and their forming mechanism were analyzed and showed that the bionic composite material exhibits almost the same reflectance spectrum with those of green leaves, and its spectrum changes little after sunlight treatment for three months, which proves a good sunlight endurance of this bionic composite material (Yang et al., 2010).

GREEN LEAF INSPIRED BIONIC PHOTOVOLTAIC PANELS

It has been observed that the silicon solar panels, in strong solar light, can heat up by 70°C, resulting in the loss of electricity generation. Contrarily, plant leaf structures possess several adaptations that allow them to sustain their temperature to 40 – 45°C in full sunlight, even when water evaporation is suppressed. Plants have adapted several strategies such as leaf size limitation, wind aerodynamics optimization, as well as efficient thermal emission of absorbed solar energy. Zahr et al (2010) analyzed the serrated structure of ordinary green leaves distributed over typical twig structures and observed Australian palm tree *Licuala ramsayi* to be a very useful bionic model. *L. ramsayi* applies optimized aerodynamic properties for cooling along with a large area for harvesting solar energy to restructure itself as a protection against strong winds.

The environmental, chemical and factors affect the plant photosynthetic process and in turn, its growth and function. These factors are further affected by nutrients and hormones, or chemical messengers which are transported throughout the plant through its vascular circuits. The artificial regulation of plant processes today can be achieved by exposing it to exogenously added chemicals or through genetic tools that can alter the metabolism and signal transduction pathways in desired ways.

NANOPARTICLES HELP IN FIGHTING THE DISEASES AND CAN ACT AS NANO- PESTICIDE AND HERBICIDE

- Several studies have reported that nanoparticles can be used to suppress pathogens which increased crop growth by reducing 50% colony formation of pathogenic fungi that caused disease in ryegrass. For instance,
- Lamsal et al (2011) have also reported that application of Ag NPs enhanced the disease suppression. Combined activities of Ag NP with the fungicide flucanazole were found to be effective against *Candida albicans*, followed by *Phoma glomerata* and *Trichoderma* sps.

- ZnO NPs reduced growth by 26% of Fusarium graminearum grown in mungbean broth agar.
- MgO NPs exhibited significant antimicrobial activity due to strong interaction with a negative surface of bacterial membranes.
- Chemically synthesized Cu NPs demonstrated higher pathogenic fungal inhibition in comparison to the fungicide bavistin.

The problem pests and herbs can be significantly resolved by using nanoparticles, which reduces the amount of herbicides requirement for removal of weeds (reviewed by Singha et al., 2015). Gruere et al (2011) have reported that with the smart delivery system and active ingredients, based on the soil condition, herbicides are released.

Conventional methods of pest and pathogen control have adverse effects on both the environment and economy. Therefore, the nanoparticle applications are attracting significant interest for environmentally and economically sustainable agriculture. The increase in crop growth/yield may be the result of reduced disease presence, which is possible either directly due to the nanoparticle anti-pathogenic activity, or may be through the induction of key defensive pathways in the plant (Bhattacharyya et al., 2010). Silver nanoparticle of *Tinospora cordifolia* which have been biosynthesised, showed maximum cidal effect against the head louse *Pediculus humanus* and fourth instar larvae of *Culex quinquefasciatus* and *Anopheles subpictus* (Jayaseelan et al., 2011).

Nanomaterilas as nanofertilizers can provide nutrients to the plants or they assist to augment the activities of conventional fertilizers. Replacement of nanofertilizers for traditional fertilizer is beneficial as its application is to release nutrients into the soil steadily and in a controlled way, thus reducing the frequency of application and preventing environmental pollution. The nano-organic iron chelated fertilizers have been shown to increase absorption, photosynthesis and leaf surface area expansion (Reviewed by Singha et al., 2015).

ROLE OF NANOPARTICALS IN PHOTOSYNTHESIS

Researchers at MIT, USA have reported that the plants' ability to capture light energy can be increased by 30 percent by embedding carbon nanotubes in the chloroplast of their leaves. New functions can also be imparted to these plants by incorporating nanomaterials in them.

Researchers have tried to (Giraldo et al., 2014) improve ability of the plants to capture light energy by 30 percent by embedding carbon nanotubes in the chloroplast, the plant organelle where photosynthesis takes place. Using another type of carbon nanotube, they also modified plants to detect the gas nitric oxide.

Dr Strano and his co-workers stumbled on this idea of nanobionic while working on plant cell-inspired self-repairing solar cells. His primary aim was to enhance the photosynthetic function of chloroplasts isolated from plants, to be used later in solar cells.

Chloroplasts house all the machinery needed for photosynthesis. The first phase starts with absorption of light by chlorophyll, followed by excitation of electrons that flow through the thylakoid membranes of the chloroplast. This electrical energy is captured by the plant and is used to power the manufacture of sugars in the second stage of photosynthesis.

Even after being removed from plants, chloroplasts can still perform these reactions. However, they start to break down, after a few hours because of the damage caused to the photosynthetic proteins by light and oxygen. While this kind of damage is repaired by the plants, the extracted chloroplasts can't do their own repair.

To enhance the productivity of the chloroplasts, cerium oxide nanoparticles, also known as nanoceria were embedded in them using a new technique they developed called lipid exchange envelope penetration, or LEEP. These particles scavenge oxygen radicals as they are very strong antioxidants and protect the chloroplasts from damage. In these chloroplasts, levels of damaging molecules dropped dramatically.

Moreover, the researchers also embedded semiconducting negatively charged DNA coated carbon nanotubes, into the chloroplasts through same delivery technique. The carbon nanotubes could function as artificial antennae and act as a "prosthetic photoabsorber," to allow chloroplasts to capture wavelengths of light not in their normal range, such as ultraviolet, green, and near-infrared, and enhance the productivity of the plants because typically plants make use of only about 10 percent of the sunlight available to them.

The photosynthetic activity was 49 percent greater (as measured by the rate of electron flow through the thylakoid membranes) than that in isolated chloroplasts that did not have nanotubes embedded in them. It was also observed that the chloroplasts remained active for a few extra hours when nanoceria and carbon nanotubes were delivered together.

NANOBIOSENSORS: THE "BOMB DETECTOR" BIONIC PLANTS

Nanosensors dispersed in the field can detect the presence of plant viruses, other crop pathogens and the level of soil nutrients. Mousavi and Rezaei (2011) reported that nano smart dust i.e. tiny wireless sensors and transponders can evaluate environmental pollution quickly. Insects, disease, pathogens, chemicals and contaminants can be rapidly detected by portable nano devices and results in faster treatment. Nanotechnology based devices will increase the use of sensors for real time monitoring of crops.

The researchers also showed that *Arabidopsis thaliana* plants could be turned into chemical sensors because of carbon nanotubes that detect the gas nitric oxide, an environmental pollutant produced by combustion (Giraldo et al., 2014).

The researchers hope that they will be able to develop plants by adapting different target sensors that could be used to monitor environmental pollution, toxins, pesticides, or fungal infections.

The nanoparticles and carbon nanotubes (tiny cylinders of carbon) were implanted into the leaves of the spinach plant and the nitro-aromatics delivered into the water taken up by the roots and directly to the leaves in droplets. In nearly 10 minutes, the timetaken by the water to finally reach the tip, the researchers shine a laser onto the leaf to read the signal that comes out as near-infrared fluorescent light emitted by the embedded nanotubes. The signal is detected by a small infrared camera connected to a small, cheap Raspberry Pi computer, or, with a smartphone.

Prof Strano's lab has previously developed carbon nanotubes that can be used as sensors to detect hydrogen peroxide, TNT, and the nerve gas sarin, and the signals are read in the form of emitted rays. These bionic plants could find use in defence applications, and for monitoring terrorism related activities.

A FEW ISSUES

Nanosensors offer the advantage of being small, portable, sensitive and they are accurate, reproducible and stable too, thus they can replace the conventional sensors. Soil temperature and moisture are very important indicators and play significant role in agriculture. With the help of microelectronic circuit, biosensors are able to sense and monitor the temperature and moisture of the soil (Sharon et al., 2010). It is now clear that nanobiosensors can be effectively used for sensing a wide variety of fertilizers, herbicides, pesticides, insecticides, pathogens and thus can support sustainable agriculture by enhancing crop productivity.

In the study of phytotoxicity, uptake and accumulation of nanoparticles, we are still in the nascent stage. Although preliminary results of the activities of NPs in plant species are there, the differences in responses of plant species remains unexplored. The genetic response of the bionic plants with respect to different vascular structure is also a topic of discussion. More studies are required in the field to confirm several hypotheses that have been proposed. The concern on the toxicity of the nanoparticles in the environment has emerged during the last decade, with several-fold increase in the applications these particles.

Finally, these techniques may create unforeseen problems, are may have potential risks. For example, the spread of pollen from these bionic plants to natural populations could be a potential problem and may be warranted, in addition to changes in crop production patterns.

SUMMARY

Ionic transport and conductivity are fundamental to plant physiology. In vivo electronic functionality can be added to plants successfully, as is reported in *R. floribunda* plants. These findings pave the way for organic electronics in plants. Scientists have identified delivery and sensor devices as an interesting concept of e-Plants for feedback-regulated control of plant physiology. These controls could possibly complement the plant molecular genetic techniques and agriculture, broadly. The approach to take examples from the plant kingdom and to translate specific features in artificial solutions, especially in the robotic field, is gaining momentum.

REFERENCES

1. Almeelbi T, Bezbaruah A (2014) Nanoparticle-sorbed phosphate: iron and phosphate bioavailability studies with Spinacia oleracea and Selenastrum capricornutum. Acs Sustain Chem Eng 2(7), 1625–1632
2. Awwad AM, Albiss BA, Salem NM (2015) Antibacterial activity of synthesized copper oxide nanoparticles using Malva sylvestris leaf extract. SMU Medical Journal 2(1), 91-101.
3. Barthlott W, Neinhuis C (1997) Purity of the sacred lotus, or escape from contamination in biological surfaces. Planta 202, 1–8. doi:10.4315/0362-028X. JFP-12-021
4. Bhattacharyya A, Bhaumik A, Rani PU, Mandal S & Epidi TT (2010) Nanoparticles – a recent approach to insect pest control. Afr J Biotechnol 9(24), 3489–3493.
5. Bhushan B (2009) Biomimetics: lessons from nature-an overview. Philos Trans A Math Phys Eng Sci 367, 1445–1486. doi:10.1098/rsta.2009.0011
6. Burgert I, Fratzl P (2009) Actuation systems in plants as prototypes for bioinspired devices. Phil Trans R Soc Lond A 367, 1541–1557. doi:10.1098/rsta.2009. 0003
7. Canas JE, Long M, Nations S, Vadan R, Dai L & Luo M et al (2008) Effects of functionalized and nonfunctionalized singlewalled carbon nanotubes on root elongation of select crop species. Environ Toxicol Chem 27, 1922–1931.
8. Fischer F, Thielen M, Loprang RR, Seidel R, Fleck C, Speck T, et al (2010) Pummelos as concept generators for biomimetically inspired low weight structures with excellent damping properties. Adv Eng Mater 12, B658–B663. doi:10.1002/adem.201080065
9. Forterre Y (2013) Slow, fast and furious: understanding the physics

of plant movements. J Exp Bot 64, 4745–4760.doi:10.1093/jxb/ert230

10. Fratzl P, Barth FG (2009) Biomaterial systems for mechanosensing and actuation. Nature 462, 442–448.doi:10.1038/nature08603
11. Giraldo JP, Landry MP, Faltermeier SM, McNicholas TP, Iverson NM, Boghossian AA, et al (2014) Plant nanobionics approach to augment photosynthesis and biochemical sensing. Nature Materials 13, 400–408.
12. Gruère G, Clare N, Linda A (2011) Agricultural, Food and Water Nanotechnologies for the Poor Opportunities, Constraints and Role of the Consultative Group on International Agricultural Research. J Int Food Policy Res Inst. 1-35.
13. Jayaseelan C, Rahuman AA, Rajakumar G, et al (2011) Synthesis of pediculocidal and larvicidal silver nanoparticles by leaf extract from heartleaf moonseed plant, Tinospora cordifolia Miers. Parasitol Res 109(1), 185–194.
14. Lahiani MH, Dervishi E, Chen J, Nima Z, Gaume A, Biris AS & Khodakovskaya MV (2013) Impact of carbon nanotube exposure to seeds of valuable crops, ACS Appl Mate 5, 7965–7973.
15. Lamsal K, Kim SW, Jung JH, Kim YS, Kim KS, Lee YS (2011a) Application of silver nanoparticles for the control of Colletotrichum species in vitro and pepper anthracnose disease in field. Mycobiology 39(3), 194–199.
16. Martone PT, Boller M, Burgert I, Dumais J, Edwards J, Mach K, et al (2010) Mechanics without muscle: biomechanical inspiration from the plant world. Integr Comp Biol 50, 888–907.doi:10.1093/icb/icq122
17. Meyers MA, Chen P-Y, Lin AY-M, Seki Y (2008) Biological materials: structure and mechanical properties. Prog Mater Sci 53, 1–206.doi:10.1016/j. pmatsci.2007.05.002
18. Mousavi SR, Rezaei M (2011) Nanotechnology in agriculture and food production, J Appl Environ Biol Sci 1(10), 414– 419
19. Nair R, Varghese SH, Nair BG, Maekawa T, Yoshida Y, Kumar DS (2010) Nanoparticulate material delivery to plants. Plant Sci 179, 154–163.

20. Rajakumar G, Rahuman A A, Priyamvada B, Khanna V G, Kumar D K, Sujin PJ (2012) Eclipta prostrata leaf aqueous extract mediated synthesis of titanium dioxide nanoparticles. Mater Lett 68, 115–117.
21. Rosa De, Monreal MC, Schnitzer C Walsh M R, Sultan Y (2010) Nanotechnology in fertilizers. Nat Nanotechnol 5, 91.
22. Sharon M, Choudhary AK, Kumar R (2010) Nanotechnology in Agricultural Diseases. J Phytol 2(2010) 83-92.
23. Singha A, Singha NB, Hussaina I, Singha H, Singh SC (2015) Plant-nanoparticle interaction: An approach to improve agricultural practices and plant productivity. Int J Pharm Sci Invent 4(8), 25-40.
24. Stavrinidou E, Gabrielsson R, Gomez E, Crispin X, Nilsson O, Simon DT, Berggren M (2015) Electronic plants. Sci Adv 1:e1501136 20
25. Thielen M, Schmitt CNZ, Eckert S, Speck T, Seidel R (2013) Structure– function relationship of the foam-like pomelopeel (Citrusmaxima) –an inspiration for the development of biomimetic damping materials with high energy dissipation. Bioinspir Biomim 8, 025001.doi:10.1088/1748-3182/8/2/025001
26. Tong J, Xu S, Chen D, Li M (2017) Design of a Bionic Blade for Vegetable Chopper. J Bionic Eng 14 (1), 163–171.
27. Velcro SA(1955) Improvements in o relating to a Method and a Device for Producing a Velvet Type Fabric. PatentNo.721338Switzerland.
28. Wang S, Ren L, Liu Y, Han Z, Yang Y (2010) Mechanical Characteristics of Typical Plant Leaves. J Bionic Eng 7(3), 294-300.
29. Wang X, Han H, Liu X, Gu X, Chen K & Lu D (2012) Multi-walled carbon nanotubes can enhance root elongation of wheat (Triticum aestivum) plants. J Nanopart Res 14(6), 1–10.
30. Wong MH, Giraldo JP, Kwak S-Y, Koman VB, Sinclair R, Lew TTS, et al (2017) Nitroaromatic detection and infrared communication from wild-type plants using plant nanobionics. Nat Materials 16, 264–272, doi:10.1038/nmat4771
31. Yang F, Hong F, You W, Liu C, Gao F, Wu C, Yang P (2006) Influence of nano-anatase TiO2 on the nitrogen metabolism of growing spinach. Biol Trace Elem Res 110(2), 179–190.
32. Yang Y, Liu Z, Hu B, Man Y, Wu W (2010) Bionic Composite

Material Simulating the Optical Spectra of Plant Leaves. J Bionic Eng 7, S43-S49.

33. Zähr M, Friedrich D, Kloth TY, Goldmann G, Tributsch H (2017). Bionic Photovoltaic Panels Bio-Inspired by Green Leaves. J Bionic Eng 7(3), 284-293.

CHAPTER 8

APPLICATIONS AND ENGINEERING ANALYSIS OF LOTUS ROOTS UNDER EXTERNAL WATER PRESSURE

Yiyun Zhu[1], Chang Jiang Wang[2], and Diane Mynors[2]

[1]School of Civil Engineering and Architecture, Xian University of Technology, No. 5 Jinhua South Road, Xian, Shaanxi 710048, China

[2]School of Engineering and Informatics, University of Sussex, Brighton BN1 9QJ, UK

ABSTRACT

Engineers can learn from nature for inspirations to create new designs. The internal structure of lotus roots with several oval holes was studied in this paper for engineering inspirations. The structural performance of lotus roots under outside water pressure was simulated and compared with various cross-sectional areas. The distribution of stresses in the cross-sectional area of lotus roots was analyzed and presented. It was found that the maximum compressive stresses in the cross-sectional area of lotus roots were occurring at the long axis ends of the holes. This was very different from that of circular holes. Further analysis on the triaxiality factors revealed that the cross-sectional area of the lotus root resulted in large areas of high triaxiality factors. The resulting hydrostatic stress in the cross-sectional area of lotus root ranges from zero to 2.7

times the applied outside pressure. In contrast, the hydrostatic stress in a cylindrical cross-sectional area is a fixed value. The study showed that the lotus root and the orientation of the oval holes could be mimicked in the design of new structures, for example, underwater pipes and vessels.

INTRODUCTION

Through evolution, nature has learned to achieve maximal performance by using minimum resources. It has evolved and optimized a large number of materials and structured surfaces with rather unique characteristics [1]. Therefore, adopting designs based on the study of plants and animals in the field of biomimetics or bionics is important as biological systems produce many functions that can be applied in engineering; many examples have been presented and discussed by Vincent [2]. The benefits gained from biomimetics are not totally obvious; therefore, the practical use of mechanisms of functions in engineering and other disciplines is still young [3]. The biological system should be studied and understood before the ideas from biology can be transferred into engineering and design.

Structural optimization is very important in the design of mechanical systems in industry. Shape optimization of engineering components can follow the design rules of nature; for example, Mattheck [4] studied the tree fork and observed that trees can maintain a uniform stress distribution at their surface through load-adaptive growth. Mattheck [4] then proposed a method of tensile triangles to remove unloaded parts within a structure in order to save materials.

In this paper, lotus roots with large and small holes under external water pressures will be studied to extract nature's design principles. Lotus roots are found buried in anaerobic sediment and are characterized by having oval holes for obtaining oxygen. Mevi-schutz and Grosse [5] conducted experiments that showed that thermoosmotic gas transport could drive oxygen flow from the lotus leaves to the roots. Mevi-schutz and Grosse [6] also showed a lacunar pressure of up to 166 ± 44 Pa that could be measured in both young and old lotus leaves. The standard atmospheric pressure is 101325 Pa; therefore, it can be reasonably assumed that the gas pressures inside the lotus root holes are close to the atmospheric pressure when the structural analysis was conducted in this paper.

Dominy et al. [7] have studied the mechanical properties of plant underground storage organs. They found that rhizomes were the most resistant to deformation and fracture, followed by tubers, corms, and bulbs. They used a portable universal tester to estimate Young's modulus and fracture toughness of a range of plant species, with Young's modulus varying between 0.8 MPa and 18.7 MPa.

Vincent [8] reported many advantages of using holes in engineering structures, for example, making an object lighter and more durable, and holes also can affect the way that a material fails. It was pointed out by Vincent [8] that engineers and designers treat holes with suspicion and are not using their advantages because we do not always know how best to use them.

The study of the effect of holes on the strain distribution in Campaniform Sensilla by Vincent et al. [9] showed that the orientation of the hole with respect to the applied load is significant, and the effects of grouping and mutual proximity of the holes are important in strain magnification as well.

The lotus root has a unique geometry with its canals regularly aligned. Through the study of the lotus root's porosity and orderly arranged pores, the lotus root has already provided engineering inspirations for the designs of a multibore hollow fiber membrane [10] and a porous nanocomposite polymer electrolyte [11]. It has also been applied to the development of porous carbon steels [12].

Chen and Zhang [13] reported that the enlargement of parenchymatous cells resulted in the growth or thickening of the rhizome. Niklas [14] reported that tissue composite modulus should be named for the elastic modulus obtained from mechanical test, because it is different from the modulus for solid materials. The elastic modulus of parenchyma tissue is reported to be between 3 MPa and 6 MPa; the compressive strength is between 0.27 MPa and 1.3 MPa [15]. Stresses will be developed in the lotus roots when outside water/mud loads are applied; these internal stress states can affect cell expansion. To analyze the state of stress in lotus roots, triaxiality and hydrostatic stress will be discussed.

Material properties can be affected by hydrostatic stress in material deformations. Triaxiality is mainly used to describe the forming limit of materials and ductile fracture criteria. The triaxiality factor (TF) in a material is a ratio of the hydrostatic stress and the von Mises stress resulted from loading. Bridgman [16] reported that hydrostatic stress could increase the ductility of metals but not result in plastic deformation. With certain hydrostatic stress levels, brittle materials can be deformed like ductile materials. In uniaxial tension, the TF is 1/3; in uniaxial compression, TF is −1/3; and, in biaxial tension, TF is 2/3. When compressive principal stresses occurred in materials, TF is less than 1/3. Kweon [17] studied material damage and showed that no shear damage can occur when the TF is smaller than −1/3. Bao and Wierzbicki [18] showed that fracture never occurs when the stress triaxiality value is smaller than −1/3.

The novelty of this paper is to analyze and reveal the effects of the layout and orientation of the large and small holes on the stress distribution in the cross-sectional area of lotus roots through finite element simulations. To this end, the hydrostatic stress and triaxiality in the cross-sectional area of lotus with oval holes will be analyzed and compared with the cross-sectional areas of various artificial arrangements of holes. From this, the natural design principles of the lotus root is extracted and applied to solve an engineering problem as a proof of concept.

METHODS

The Structure of Lotus Roots

The fresh lotus root shown in Figure 1(a) was bought from a supermarket and the origin is not known, only the dimensions of the cross-sectional area are important in this research. The lotus root was cut at the middle section using a kitchen knife and the slice used for study is shown in Figure 1(b). The slice dimensions were measured using the pixels alongside the scale of the picture in Figure 1(b), and the holes were numbered and labelled for 2D finite element modelling and engineering analysis.

Figure 1: (a) Lotus roots and (b) a slice of lotus root used for the analysis.

The area of each hole in the lotus root shown in Figure 1(b) is included in Table 1. The solid area of the cross-sectional area of the lotus root in Figure 1(b) is 2138 mm^2. The total area of large holes numbered from 1 to 10 is 506 mm^2, with a mean value of 50.6 mm^2 and a standard deviation of 26.7 mm^2. The total area of the small holes labelled alphabetically from a to k is 28 mm^2, with a mean value of 2.5 mm^2 and a standard deviation of 1.1 mm^2. The total area of all the holes and the solid area is 2672 mm^2. Therefore, the porosity of the cross-sectional area of the lotus root is 20%.

Table 1: Area of holes in the cross-sectional area of the lotus root.

Large hole	1	2	3	4	5	6	7	8	9	10	
Area (mm^2)	90.8	48.8	58.5	21.9	14.4	55.6	59.8	85.1	56.6	14.9	
Small hole	a	b	c	d	e	f	g	h	i	j	k
Area (mm^2)	2.1	2.1	3.2	2.8	3.0	2.9	4.5	1.3	1.2	3.8	1.1

Finite Element Model of the Cross-Sectional Area of Lotus Root

ANSYS [19] finite element software was used for the simulation. The finite element model of the cross-sectional area of the lotus root and the meshes are shown in Figure 2. The element size of 0.2 mm is used after convergence testing; the finite element model consists of 46733 2D quadrilateral elements. A plane strain analysis is used for the stress analysis and finite element modelling. The specific lotus root used in this research was not mechanically tested, as it is difficult to characterize the mechanical properties of biomaterials. However, for comparative analysis, assumptions from prior studies can be reasonably used: an elastic modulus of 5 MPa is adopted from the work by Gibson [15] and

Poisson's ratio of 0.3 was used from the work by Wei et al. [20] and Hamza et al. [21]. An outside pressure of 0.01 MPa was used to simulate a water depth of 1 meter, and this pressure is applied to the external boundary of the cross-sectional area of lotus root. Zero displacements are applied at a node on the center hole of section to provide numerical stability during solving process.

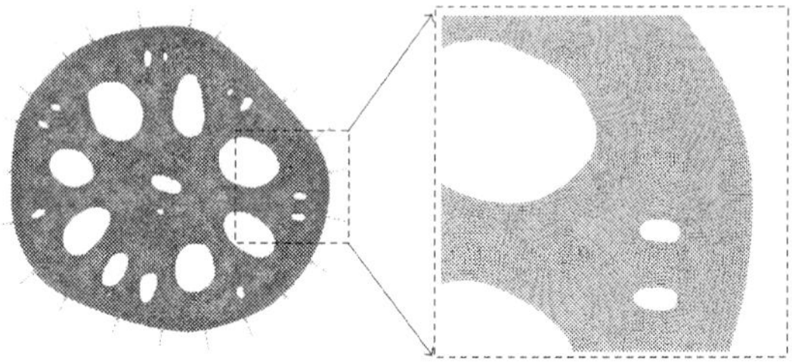

Figure 2: Finite element model of the cross-sectional area of lotus root.

Stresses in the Thick-Walled Cylinder

When the cross-sectional area of the lotus root is equivalently represented as a cylindrical cross section as shown in Figure 3, the outer and inner diameters of the hollow circular cross section are 58.3 mm and 26.1 mm in order to maintain the same values of porosity and area. Using Lamé's equation (1) for thick-walled cylinders, the tangential compressive stress on the inner surface of the hollow circular cross-sectional area is 0.025 MPa when the area is subject to an outside water pressure of 0.01 MPa:

$$\sigma_\theta = \frac{-P_o r_o^2}{r_o^2 - r_i^2}\left(1 + \frac{r_i^2}{r^2}\right). \quad (1)$$

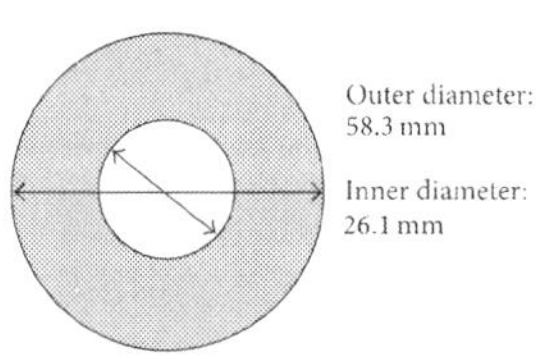

Figure 3: A cylindrical cross section of the same area and porosity of the lotus root.

RESULTS AND DISCUSSION

Compressive Stresses in the Cross-Sectional Area of the Lotus Root

To understand the influence of hole shape, the actual holes in the cross-sectional area were replaced with 10 circular holes for comparison. Each circular hole has a diameter of 4.12 mm to create the same porosity in the modified cross sections when compared to the actual lotus root. One circular hole is located at the center of the lotus root cross section and the other nine circular holes are equally distributed along the circumference of a "planetary" circle with its center coinciding with the center of the lotus root. The locations of the nine outer circular holes are varied by changing the diameter of the planetary circle and the effects on stress distributions are studied. The diameter of the planetary circle is varied from 22 mm, 20 mm, 17 mm, and 14 mm to 13 mm. These five cases are labelled as (a) to (e) as shown in Figures 4(a) to 4(e) respectively.

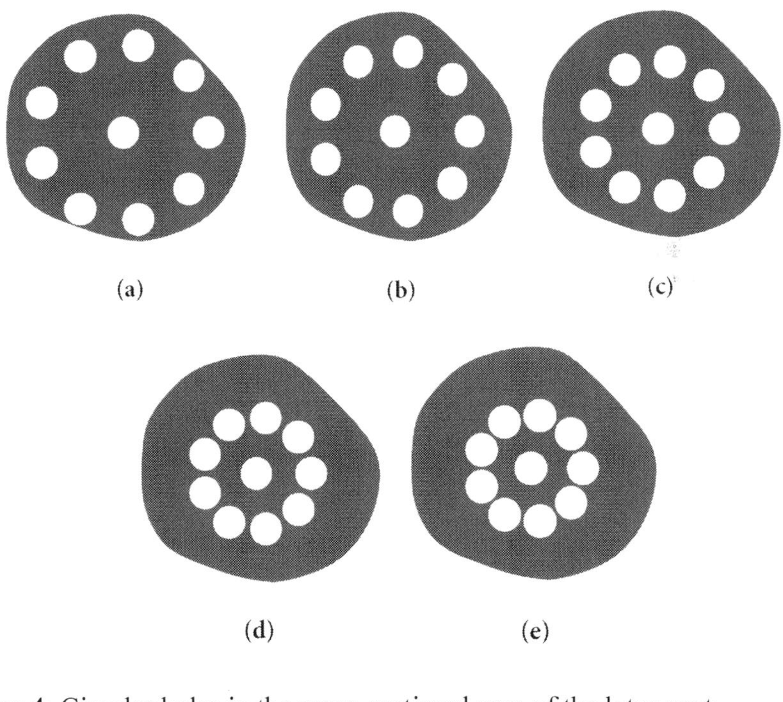

Figure 4: Circular holes in the cross-sectional area of the lotus root.

In Figure 5(a), the highest compressive stresses were developed in the area of small holes. It was noted that the maximum compressive stress in the cross-sectional area of the lotus root with smaller holes (case 1) was 0.058 MPa and 0.060 MPa in the cross-sectional area without smaller holes (case 2). These compressive stresses are also greater than that in the cylindrical cross-sectional area shown in Figure 3. Comparing the stresses in Figures 5(a) and 5(b), the maximum compressive stress in the lotus root with the small holes was reduced by 0.002 MPa, yielding a reduction of 3%. These small holes reduce the stress in the cross-sectional area of lotus root and therefore they should be included in the structural analysis.

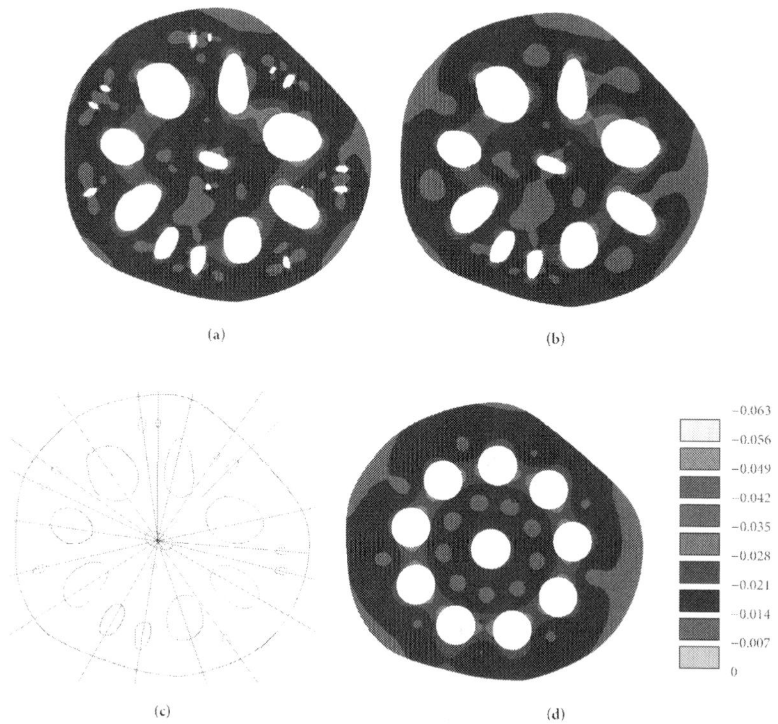

Figure 5: Compressive stress (MPa) in the lotus root (a) with small holes, (b) no small holes, (c) radial lines through oval holes, and (d) compressive stress in the cross-sectional area with circular holes.

From the compressive stress plot in Figure 5(a), it can be seen that the compressive stresses are highest at the two major axis ends of the oval holes. It was noticed that the compressive stresses are located mostly at where the oval holes intersect with the radial lines that are shown in Figure 5(c) in the cross-sectional area of lotus root. However, the compressive stress distributions around circular holes are not the same as the oval holes. The compressive stresses are located near where the circular holes intersect with the planetary circle, as shown in Figure 5(d).

Tensile Stresses in the Cross-Sectional Area of the Lotus Root

Tensile stresses in the cross-sectional area of the lotus root are presented in Figure 6 where the maximum tensile stress value is 0.00063 MPa, which is about 1% of the maximum compressive stress in the lotus root. The tensile stresses could be close to zero if the surfaces of the holes in the finite element model were a true representation of the natural lotus root.

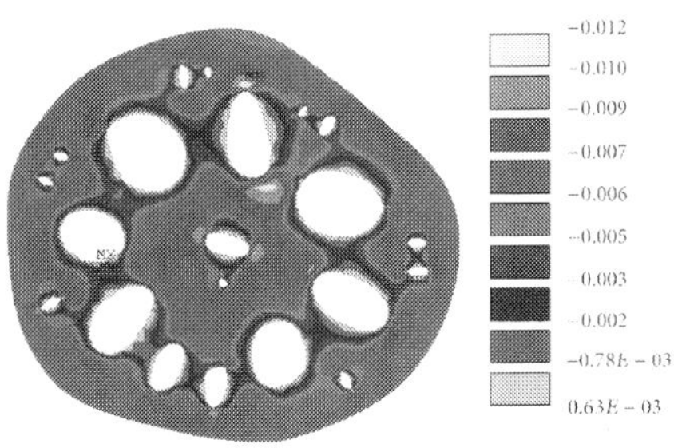

Figure 6: Tensile stress (MPa) distribution in the cross-sectional area of the lotus root.

The von Mises Stresses in the Cross-Sectional Area of the Lotus Root

The von Mises criterion or maximum distortion energy criterion is associated with the change in shape of the material and is normally applied to ductile materials like metals. However, because lotus roots are subjected to water pressure acting on the external surfaces, it is considered useful that the von Mises stress is discussed in this paper.

The von Mises stress distributions in the actual cross-sectional area of the lotus root (case 1) and with circular holes (case c) are shown in Figure 7. It was noticed that higher stresses were developed around the holes in the actual cross-sectional area of the lotus root; it was also calculated that the total area of stresses between 0 MPa and 0.006 MPa in the actual lotus root cross section is smaller than that of case c.

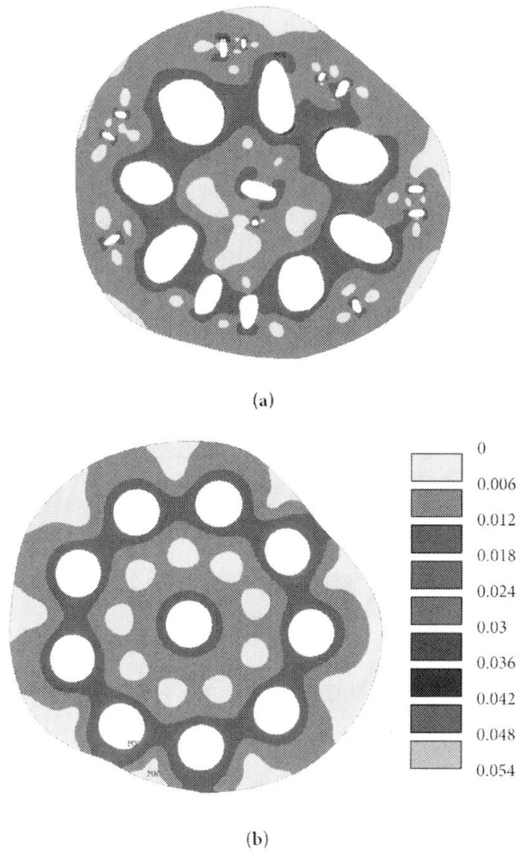

Figure 7: The von Mises stress (MPa) distribution: (a) the actual lotus root and (b) lotus root with circular holes.

In order to compare the von Mises stress distribution, von Mises stresses between 0.008 MPa and 0.012 MPa were studied, that is, 80% to 120% of the applied outside pressure. These stresses are shown in Figure 8 and it is noted that these stresses are distributed close to the holes. In order to show the effect of the distribution of the holes on the stress values and stressed area, the area of the von Mises stresses between 0.008 MPa and 0.012 MPa in each case is listed in Table 2. In case 1, the total area of having a von Mises stress between 0.008 MPa and 0.012 MPa is 691 mm^2, which is 32.3% of the solid area of the cross-sectional area of the lotus root. Comparing case 1 and case 2 in Table 2, when the small holes were considered in the modelling of the lotus root, nearly more than 10% of the areas are in stress states between 0.008 MPa and 0.012 MPa. Comparing case 1 with cases a, b, and c, it was noticed that the circular holes in cases a, b, and c resulted in a smaller area, being in the stress states between 0.008 MPa and 0.012 MPa. Cases d and e showed a larger area, being in the stress states between 0.008 MPa and 0.012 MPa. However, cases d and e are not truly representative of actual lotus root because these circular holes are very close to each other and little material is left between the holes, weakening the structure. The extreme cases of d and e showed that the lotus root should be correctly mimicked in engineering applications.

Table 2: Area in the lotus root cross section with von Mises stresses between 0.008 MPa and 0.012 MPa.

	Case 1	Case 2	Case a	Case b	Case c	Case d	Case e
Area (mm^2)	691	641	569	471	532	908	1013

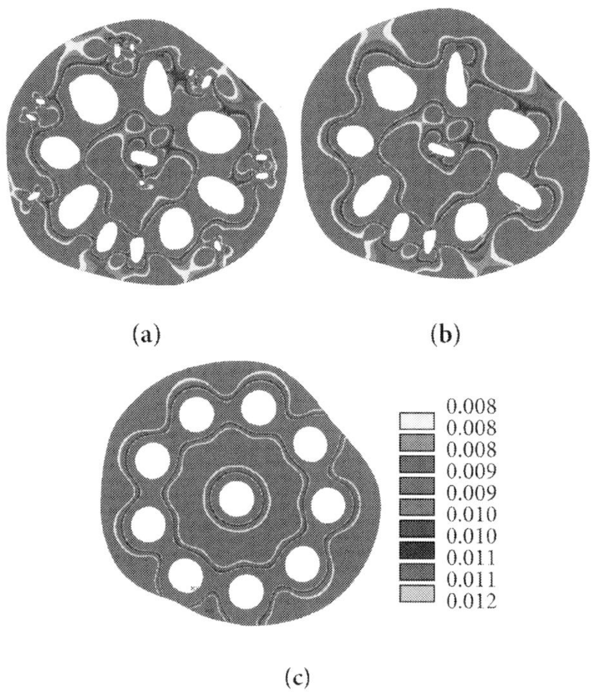

Figure 8: The von Mises stress (MPa) in the cross-sectional area of lotus root (a) with small holes, (b) no small holes, and (c) with circular holes.

Triaxiality Factors in the Cross-Sectional Area of Lotus Root

Influence of holes on the triaxiality factors in the cross-sectional area shown in Figure 2 will be compared with the cylindrical cross-sectional area shown in Figure 3. The theoretical triaxiality factors in the cylindrical cross-sectional area will be calculated and used for analysis.

With an outside water pressure load of 0.01 MPa, the triaxiality factors in the cross-sectional area of the lotus root and the cylindrical cross-sectional area is shown in Figure 9. It can be seen that high triaxiality factors are at the vicinity of these holes. For a cylindrical cross-sectional area, the triaxiality factors under plane strain assumption is shown by (2), where r, r_i, and ν are the outside, inside radius and Poisson's ratio, respectively:

$$\mathrm{TF} = \frac{2(1+\nu)}{\sqrt[3]{3\left(r_i^4/r^4\right)+(1-2\nu)^2}}. \qquad (2)$$

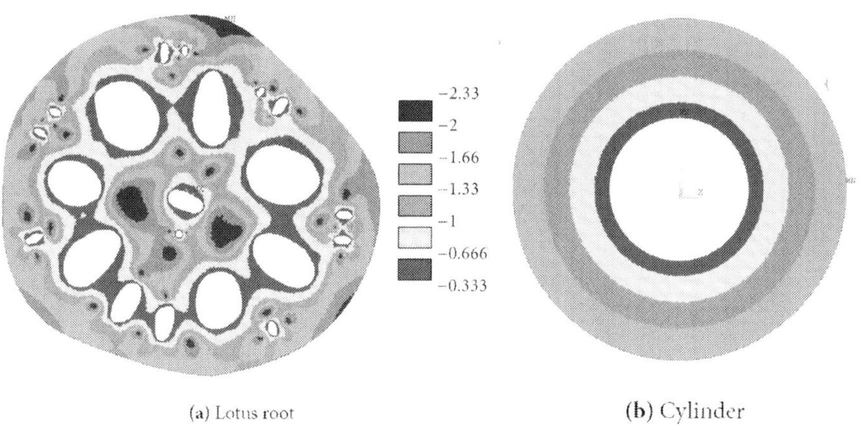

Figure 9: Triaxiality factors in the cross-sectional areas.

Taking ν as 0.3, since $r_i/r \approx 1$, for a thin walled circular crosssectional area, then TF = −0.49, when $r_i = 0$, TF = −2.17.

In order to show the difference between triaxiality factors in the cross-sectional areas shown in Figure 9, the area of each triaxiality factor group is calculated and shown in Table 3. Elements of triaxiality factors from −1/3 to −2/3 are collected as group 1, which represents the TF value close to a thin walled cylindrical cross-sectional area. Elements of triaxiality factors from −2 to −7/3 are collected as group 6, which represents the value of a solid circular cross-sectional area. From the analysis of triaxiality factors in the cross-sectional area of lotus root and hollow cylinders, there are large amounts of areas with triaxiality factors between −1/3 and −2/3 in the cross-sectional area of the lotus root compared to that of cylindrical cross-sectional area, that is, showing the behavior of a thin walled section when the lotus root is under outside water pressure.

Table 3: Area of groups of triaxiality factor (mm^2).

Group of triaxiality factor	1	2	3	4	5	6
Triaxiality factor	−1/3 to −2/3	−2/3 to −1	−1 to −4/3	−4/3 to −5/3	−5/3 to −2	−2 to −7/3
Lotus root section	330	443	440	586	261	79
Cylindrical section	214	456	604	861	0	0

Hydrostatic Stress to External Pressure Factor (HPF)

To compare and analyze hydrostatic stress, an internal hydrostatic stress pressure factor (HPF) is introduced here; it is the hydrostatic stress developed in the cross-sectional area divided by the applied outside pressure. For example, the HPF in the cylindrical cross-sectional area shown in Figure 3 is calculated as 1.08, while the HPF in a solid circular cross-sectional area is 0.87. Analyzing the finite element numerical results, the HPF from 0.02 to 2.52 in the cross-sectional area of lotus root is shown in Figure 10(a); this is different from the cylindrical cross-sectional area that has a constant HPF value. In Figure 10(b), the HPF in the majority of the cross-sectional area of the lotus root is smaller than 1.08, and higher HPF values occur near the major axis ends of the holes in the lotus root.

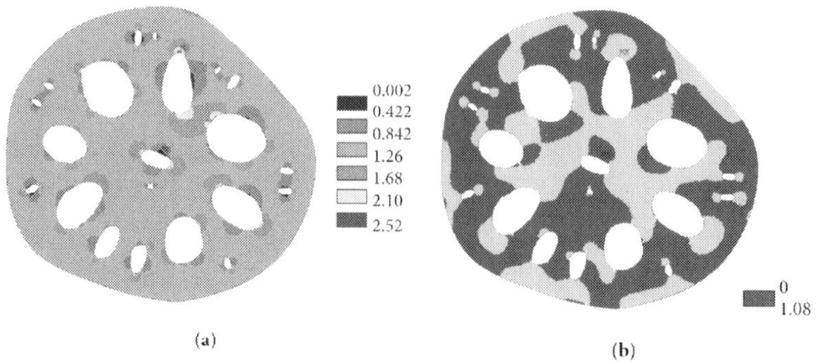

Figure 10: (a) HPF in the cross-sectional area of the lotus root and (b) HPF values less than 1.08.

Engineering Applications

Lotus roots are normally subjected to outside water pressures, and the formation of oval/elliptical holes in cross-sectional area is interesting to engineers; this research work can provide an inspiration for design ideas of underwater vessels and other structures having several compartments. To illustrate the advantage of using oval holes like that in the lotus roots, von Mises stress distributions of two cross-sectional areas are shown in Figures 11(a) and 11(b). The two cross-sectional areas have the same porosity, solid area, outer diameter, and number of holes. The von Mises stress in the section of circular holes is about 60% higher than that in the section of oval holes subject to an outside pressure of 0.01 MPa. A challenging problem in exploiting oil resources from deep seas in the petroleum industry is that the transportation pipes are subjected to high hydrostatic pressure and low temperatures. Bouchonneau et al. [22] tested an insulated deep sea pipe structure composed of a steel pipe of internal diameter 180 mm, wall thickness 12 mm, and a 5-layer insulating coating which is mainly polypropylene with a total thickness of 61 mm. Based on these dimensions and from the stress distributions shown in Figures 11(a) and 11(b), a pipe as shown in Figure 11(c) could be designed and used in the deep sea. This pipe with oval holes could be made through the extrusion of materials. The results in this research support the findings of Vincent [8] that if holes are sensibly placed and have the right dimensions, they can improve the durability of a material or structure. The hydrostatic stress and triaxiality in the cross-sectional area of the lotus roots and cross-sectional area of the pipe are different when they are under external pressure. The lotus root could inspire engineers to create new materials and novel structures.

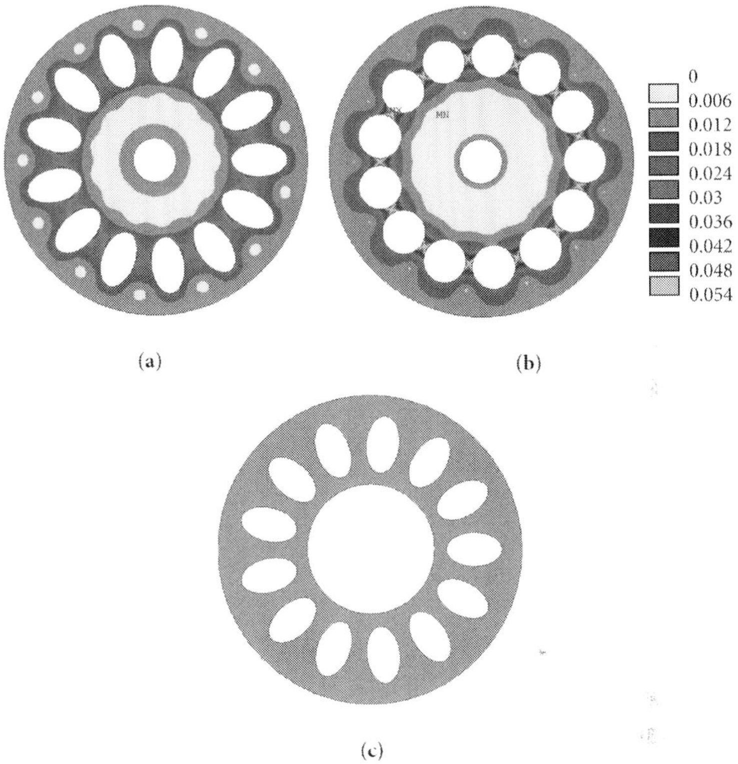

Figure 11: (a) The von Mises stress (MPa) in a section with oval holes, (b) von Mises stress in a section with circular holes, and (c) a proposed pipe cross section.

CONCLUSIONS

Because of the oval holes in lotus roots, large compressive stresses in the cross-sectional area of the lotus root occurred at the major axis ends of the actual holes. When the oval holes were simulated as circular holes, the maximum compressive stress locations were changed to where the short axis of the oval holes was located. Under the outside water pressure, the oval holes in the cross-sectional area of the lotus roots resulted in more areas having the von Mises stress close to the outside water pressure than the circular holes. Very small or negligible tensile stresses occurred in the cross-sectional areas of lotus roots under outside water pressures.

Under the outside water pressure, the cross-sectional area of the lotus root showed that the hydrostatic stress is varying. This is different from the constant hydrostatic stress values in the cross-sectional area of cylinder. The analysis also showed that large areas of high triaxiality factors occurred in the cross-sectional area of the lotus root, and all the triaxiality factors are negative. Because the shape of holes in the lotus root can change the stress states, the creation of new materials and structures can be inspired by lotus roots.

REFERENCES

1. B. Bhushan, "Biomimetics: Lessons from nature—an overview," Philosophical Transactions of the Royal Society A: Mathematical, Physical and Engineering Sciences, vol. 367, no. 1893, pp. 1445–1486, 2009.
2. J. F. V. Vincent, "Applications—influence of Biology on Engineering," Journal of Bionic Engineering, vol. 3, no. 3, pp. 161–177, 2006.
3. J. F. V. Vincent, O. A. Bogatyreva, N. R. Bogatyrev, A. Bowyer, and A.-K. Pahl, "Biomimetics: its practice and theory," Journal of the Royal Society Interface, vol. 3, no. 9, pp. 471–482, 2006.
4. C. Mattheck, "Teacher tree: the evolution of notch shape optimization from complex to simple," Engineering Fracture Mechanics, vol. 73, no. 12, pp. 1732–1742, 2006.
5. J. Mevi-schutz and W. Grosse, "The importance of water vapour for the circulating air flow through Nelumbo nucifera," Journal of Experimental Botany, vol. 39, no. 9, pp. 1231–1236, 1988.
6. J. Mevi-Schutz and W. Grosse, "A two-way gas transport system in Nelumbo nucifera," Plant, Cell & Environment, vol. 11, no. 1, pp. 27–34, 1988.
7. N. J. Dominy, E. R. Vogel, J. D. Yeakel, P. Constantino, and P. W. Lucas, "Mechanical properties of plant underground storage organs

and implications for dietary models of early hominins," Evolutionary Biology, vol. 35, no. 3, pp. 159–175, 2008.

8. J. F. V. Vincent, "Unusual uses of holes-with input from biology," Journal of the Mechanical Behavior of Biomedical Materials, vol. 4, no. 5, pp. 682–687, 2011.
9. J. F. V. Vincent, S. E. Clift, and C. Menon, "Biomimetics of campaniform sensilla: measuring strain from the deformation of holes," Journal of Bionic Engineering, vol. 4, no. 2, pp. 63–76, 2007.
10. P. Wang and T.-S. Chung, "Design and fabrication of lotus-root-like multi-bore hollow fiber membrane for direct contact membrane distillation," Journal of Membrane Science, vol. 421-422, pp. 361–374, 2012.
11. Z. H. Li, P. Zhang, H. P. Zhang, Y. P. Wu, and X. D. Zhou, "A lotus root-like porous nanocomposite polymer electrolyte," Electrochemistry Communications, vol. 10, no. 5, pp. 791–794, 2008.
12. M. Kashihara, S. K. Hyun, H. Yonetani, T. Kobi, and H. Nakajima, "Fabrication of lotus-type porous carbon steel by unidirectional solidification in nitrogen atmosphere," Scripta Materialia, vol. 54, no. 4, pp. 509–512, 2006.
13. W. P. Chen and S. M. Zhang, "The structure, elongation and thickening of rhizome in Nelumbo Nucifera Gaertn," Acta Botanica Sinica, vol. 31, no. 3, pp. 191–197, 1989.
14. K. J. Niklas, "Mechanical behavior of plant tissues as inferred from the theory of pressurized cellular solids," American Journal of Botany, vol. 76, no. 6, pp. 929–937, 1989.
15. L. J. Gibson, "The hierarchical structure and mechanics of plant materials," Journal of the Royal Society Interface, vol. 9, no. 76, pp. 2749–2766, 2012.
16. P. W. Bridgman, Studies in Large Plastic Flow and Fracture: With Special Emphasis on the Effects of Hydrostatic Pressure, McGraw-Hill, 1952.
17. S. Kweon, "Damage at negative triaxiality," European Journal of Mechanics—A/Solids, vol. 31, no. 1, pp. 203–212, 2012.
18. Y. Bao and T. Wierzbicki, "On the cut-off value of negative triaxiality

for fracture," Engineering Fracture Mechanics, vol. 72, no. 7, pp. 1049–1069, 2005.

19. ANSYS Software, ANSYS academic research, release 16.0. ANSYS Inc., Canonsburg, Pa, USA.
20. C. Wei, P. M. Lintilhac, and J. J. Tanguay, "An insight into cell elasticity and load-bearing ability. Measurement and theory," Plant Physiology, vol. 126, no. 3, pp. 1129–1138, 2001.
21. O. Hamza, A. G. Bengough, M. F. Bransby, M. C. R. Davies, and P. D. Hallett, "Biomechanics of plant roots: estimating localised deformation with particle image velocimetry," Biosystems Engineering, vol. 94, no. 1, pp. 119–132, 2006.
22. N. Bouchonneau, V. Sauvant-Moynot, D. Choqueuse, F. Grosjean, E. Poncet, and D. Perreux, "Experimental testing and modelling of an industrial insulated pipeline for deep sea application," Journal of Petroleum Science and Engineering, vol. 73, no. 1-2, pp. 1–12, 2010.

CHAPTER 9

PHYSICS AND THE MOLECULAR REVOLUTION IN PLANT BIOLOGY: UNION NEEDED FOR MANAGING THE FUTURE

Ulrich Lüttge

Department of Biology, Technical University of Darmstadt, Schnittspahnstr. 3-5, 64287 Darmstadt, Germany

ABSTRACT

The question was asked if there is still a prominent role of biophysics in plant biology in an age when molecular biology appears to be dominating. Mathematical formation of theory is essential in systems biology, and mathematics is more inherent in biophysics than in molecular biology. A survey is made identifying and briefly characterizing fields of plant biology where approaches of biophysics remain essential. In transport at membranes electrophysiology and thermodynamics are biophysical topics. Water is a special molecule. Its transport follows the physical laws of osmosis and gradients of water potential on the background of physics of hydraulic architecture. Photobiology needs understanding of the physics of electro-magnetic radiation of quantitative nature in photosynthesis and of qualitative nature in perception by the photo-sensors cryptochromes, phototropins and phytochrome in environmental responses and development. Biophysical oscillators can play a role in biological timing by the circadian clock. Integration in the self-organization of modules, such as roots, stems and leaves, for the emergence of whole plants as unitary organisms needs storage and transport of information where physical modes of signaling are essential with cross talks between electrical and hydraulic signals and with chemical signals. Examples are gravitropism and root-shoot interactions in water relations. All of these facets of plant biophysics overlie plant molecular biology and exchange with it. It is advocated that a union of approaches of plant molecular biology and biophysics needs to be cultivated. In many cases it is already operative. In bionics biophysics is producing output for practical applications linking biology with technology. Biomimetic engineering intrinsically uses physical approaches. An extreme biophysical perspective is looking out for life in space. Sustained and increased practice of biophysics with teaching and research deserves strong encouragement.

Keywords: Bionics; clock; development; electro-physiology; photobiology; photosynthesis; self-organization; signaling; water relations

THE MOLECULAR REVOLUTION, SYSTEMS BIOLOGY AND BIOPHYSICS

Molecular biology based on remarkable advances in analytical techniques was a fundamental breakthrough in the last part of the 20th century. It has advanced biology to an extent that it does not appear exaggerated to speak of a molecular revolution. However, as it is not unusual in the history of science with new breakthrough technologies, it has thence dominated biology, including plant biology. The development of analytics with raising the demand for increasingly effective machines has led to the capacity of analytically realizing complete sets of molecular contents of living cells and organisms, such as their genes (genomics), gene activities (transcriptomics), proteins (proteomics), metabolites (metabolomics), and any other "omics". Documenting these contents has been taken to mean getting hold of entire systems, and the term systems biology was coined. However, these lists of inventories are rather descriptive.

Quantitative biology should not fall behind and theory as it is intrinsic in biophysics is needed. A union of approaches needs to be cultivated.

Bioinformatics is required for handling the enormous sets of data of the omics. Conversely, biomathematics is approaching theoretical understanding. What is a system in biology? The various compounds of the omics are modules. As such they reveal little about the structure and function of complex living systems. As already Aristotle recognized, the whole is much more than the sum of its parts. Integration of modules by self-organization leads to emergence of systems with completely new properties [42,51,52]. Certainly, the knowledge of the modules is very important. However, the phenomenon of emergence tells us that we must not restrict ourselves to the pure modularity. It is now increasingly recognized that a new view on systems biology is needed which biomathematics is developing [35,36,37,54]. Physics is more disposed to mathematics and theory and biophysics rather than molecular biology can introduce this into plant biology.

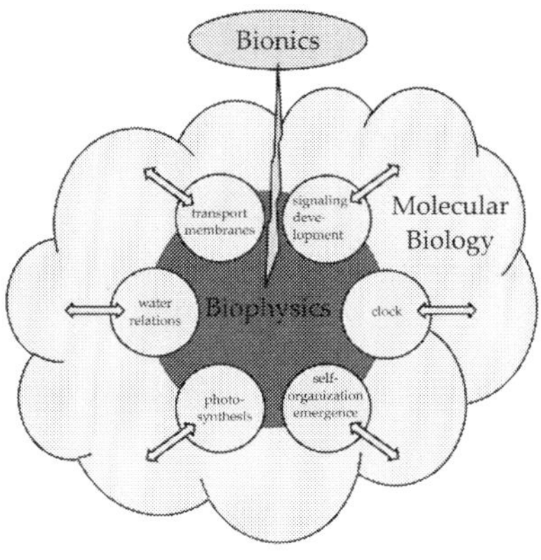

Figure 1. Facets of plant biophysics overlaid by plant molecular biology and exchanging with it, and producing output for practical applications in bionics.

Realizing that we need both, the understanding of the modules and their integration in self-organized systems, it is evident, that a union is needed for mastering future challenges. Biophysics is essential in training and research in plant biology for fulfilling the needs of further basic understanding of plant life and the practical implications given by the enormous services made by green organisms as primary producers for the existence of life on Earth. A sequence of examples is used here in an overview for developing the perspective of a union of biophysics with molecular assessments and showing where such interdisciplinary approaches are already fostering essential advances for the future. These are transport of mineral ions, water and metabolites, photobiology, development and signaling, biological timing, self-organization, and bionics, in all of which biophysical processes are inherently indispensable (Figure 1).

PLANT TRANSPORT PHYSIOLOGY

Transport is the basis of integration processes for the emergence of complex plant systems [53]. It is mediating resource allocation and distribution to subsystems. It carries signaling. Perhaps together with photobiology transport physiology is one of the two foremost topics in plant biology with implications of biophysics.

Thermodynamics and the Biophysics of Gradients

Transport across membranes is biophysically controlled by gradients between the two sides of membranes, viz. chemical gradients for electrically neutral solutes and electro-chemical gradients for electrically charged ions. Transport along chemical concentration gradients is governed by Fick's law:

$$\frac{dQ}{dt} = -D \times A \times \frac{dc}{dx} \qquad (1)$$

An amount of substance, dQ, diffuses during an amount of time, dt. A is the cross sectional area across which diffusion occurs. The concentration difference, dc, along a coordinate dx is the driving force. D is a diffusion constant. For ions deduced from thermodynamics the Nernst-equation for equilibrium and the Goldman-equation for constant electric field show quantification with gradients of concentrations or chemical activities (a) and electric potential (ΔE) across membranes [103]:

$$\text{Nernst-equation } \Delta E = -\frac{RT}{zF} \times \ln \frac{a_a}{a_b} \qquad (2)$$

where R is the universal gas constant, F the Faraday constant, z the electric ion charge and T the absolute temperature (Kelvin), subscripts a and b indicate different compartments separated by a membrane;

Goldman-equation written for three univalent ions (z = ± 1)

$$E = \frac{RT}{F} \ln \left(\frac{[K_i^+] \cdot P_K + [Na_i^+] \cdot P_{Na} + [Cl_o^-] \cdot P_{Cl}}{[K_o^+] \cdot P_K + [Na_o^+] \cdot P_{Na} + [Cl_i^-] \cdot P_{Cl}} \right) \qquad (3)$$

where ion concentrations or activities are in [], subscripts o and i refer to two sides of a membrane (e.g. outside and inside compartment) and P are the permeability coefficients for the various ions as indicated by the subscripts of element symbols.

The technical challenges right from the beginning were the chemical measurements of ion concentrations and electrophysiological measurements of membrane potential, chemistry and physics going hand in hand. The origin of plant electrophysiology and with it biophysics was possibly in 1872 when J. Burdon-Sanderson [16] measured the first action potentials of plants in the Venus' flytrap *Dionaea muscipula* after Charles Darwin had suggested it and sent him some specimen of the plant. Burdon-Sanderson used extracellular electrodes. The first recordings of electric membrane potentials of plants with intracellular electrodes were obtained only 60 years later by Umrath [101] using the large internodial cells of charophytes which allowed impalement with micro-electrodes. Further progress was obtained with other giant algal cells in addition to charophytes, e.g. *Acetabularia, Valonia, Hydrodictyon* and others [80] and only still 30 year later in the laboratory of Noë Higinbotham the first intracellular recordings of cells of higher plants were obtained [22].

Sampling for direct analysis of contents of intracellular compartments except for the vacuoles of giant algal cells was impossible. Compartmentation analysis was introduced measuring exchange of ions between tissue pre-labeled in a radioactive ion solution up to establishment of kinetic equilibrium of influx and efflux and a non-labeled solution otherwise identical to the labeling solution. A complex set of equations was developed relating fluxes and pools, i.e. for assessing the individual fluxes at the plasma membrane and tonoplast and the ion contents of cytoplasm and vacuole [78,104].

It is important to realize that now, in the age of molecular "transportomics", the compartmentation analysis with measuring kinetics of efflux of the labeled ions in exchange for the non-labeled ones are still used to biophysically quantify ionic relations [1,14,41]. Currently a huge number of membrane transporter molecules are qualitatively identified and characterized. However, molecular biology does not tell us anything about their quantitative operations, while conversely thermodynamics do not tell us anything about the nature of the molecular hardware involved. Both need each other.

Individual fluxes by specific ion transporters in complex functions of cells and whole plants are usually interconnected in functional networks and their feed-back and feed-forward dynamics. An outstanding example is the control of the opening and closing, respectively, of stomatal pores by guard cells, which is essential for the life of plants by fine-tuning their performance in the ecological starvation/desiccation dilemma of CO_2-uptake and H_2O-vapor loss, respectively. It is a biophysical problem at two different levels, namely both gas exchange and ion-transporter networks. A remarkable mathematical model for functioning of the latter has been presented by [19].

Free Energy of Molecular Pumps and Biophysics of Activity Gradients of Solutes

With thermodynamics biophysics is much involved in the debate of distinguishing passive and active transport. Passive transport is driven by existing chemical gradients of non-charged solutes or electro-chemical gradients of ions and cannot operate beyond thermodynamic equilibrium. Conversely active transport is uphill. A vigorous debate had occurred in the 1970ies over what that should imply in steady state equilibrium thermodynamics or in non-steady state dynamic irreversible thermodynamics [103]. Strictly spoken active transport had to be considered where the transport of molecules or ions (for simplicity called "particles") was directly driven by the source of metabolic energy. This was assumed to be mediated by so-called pumps, mostly ion pumps. It was termed primary active transport because it was realized that there were many uphill transports not directly linked to the ultimate metabolic energy source. These were brought about by flux couplings where the gradient of a particle *a* drove a flux of that particle which was coupled to the flux of a particle *b* which allowed the transport of particle *b* against its gradient. This was termed secondary active transport.

The concept matured to acceptance by the molecular identification of membrane proteins which functioned in primary active transport, mostly of ions. There had been the hypothesis that some primary transport processes could be directly coupled to the energy of electron transport and the redox-energy of mitochondria [44,45,46,47] or the thylakoids of chloroplasts [57,79], however, it turned out that the primary energy source was almost always ATP [53]. Proton-pumps, or H^+-ATPases, and calcium-pumps, or Ca^{2+}-ATPases, were identified at various membranes in plant cells, i.e. the plasma membrane and the tonoplast and others, and characterized in detail by methods of molecular biology. At the tonoplast an additional proton-pump using the free energy of inorganic pyrophosphate was detected, the H^+-PPase. ABC transporters (from ATP-binding casettc) are pumps directly using ATP for the transport of other solutes, e.g. glutathion-metal complexes.

The free energy available by the hydrolysis of ATP is given by

$$\Delta G_{ATP} = \Delta G^0_{ATP} + RT \ln \frac{[ADP]+[P_i]}{[ATP]} \qquad (4)$$

where ΔG^0_{ATP} is the energy under standard conditions, concentrations are in [] and P_i is inorganic phosphate. Ion pumps establish concentration gradients and electric potentials across membranes. They are electrogenic. The electrochemical ion gradient $\Delta\mu$, across a membrane separating two compartments, e.g. 1 and 2, for example for protons $\Delta\mu_{H+}$, is given by

$$\Delta\mu_{H+} = z\,F\,\Delta E_{2\text{-}1} + RT \ln\frac{c_1}{c_2} \qquad (5)$$

Biophysical assessments of such relations mark the limits within which the molecular entities of the various pumps can operate. An H^+-ATPase thermodynamically can only achieve net electrogenic H^+ pumping at $\Delta\mu_{H+} \leq \Delta G_{ATP}$.

Transporter Molecules: Setting Up Molecular Biophysics of Membranes

Molecular biology is currently providing the isolation and molecular characterization of a plethora of transporter molecules in membranes, i.e. carriers and ion channels. Among ion carriers there are cation exchangers, e.g. Na^+/H^+ and Na^+/K^+ for maintenance of Na^+-K^+-homeostasis. Among ion channels there are families of cation (K^+, Ca^{2+}) and anion (NO_3^-, Cl^-, malate^{2-}) channels [30]. One can speak of channelomics.

Techniques of molecular biology allow handling of transporter molecules, but we realize that this can only be documented and followed functionally by applying biophysical measurements:

- Messenger RNA (mRNA) and complementary DNA (cDNA) can be isolated and cloned. By introduction into host cells they can be expressed in recipients where they normally do not occur. Such heterologous expression, often in the oocytes of the frog *Xenopus laevis*, then allows studying their specific functions.
- Using gene technology transporter proteins in living plant cells can be labeled. A green fluorescent protein (GFP) is attached. Using fluorescence-optical techniques in confocal microscopy this allows sub-cellular spatio-temporal localization of the labeled proteins including specific transporter molecules in specific membrane sites.
- Transporter molecules can be modified by molecular engineering. Particular sequences of amino acids or individual amino acids of transporter proteins can be exchanged for others. In this way structure-function relationships can be assessed at the level of molecular fine-structure.

Techniques of biophysics for functional assessment of the products of molecular biology can be the traditional ones already described above. Even surface electrodes are still important devices, e.g. relating action potentials of *Dionaea muscipula* to the subsequent molecular and physiological processes at the levels of genes, enzyme proteins and resorption of solutes from the prey. This is a nice example of biophysics integrated with molecular biology [9]. However, a biophysical breakthrough was in 1976 the development of the patch-clamp technique by E. Neher and B. Sakmann (Nobel Prize 1991) [71,72,81] originally for animal cells and in 1984 with the first measurements in plant cells [85].

- A small piece of a membrane ("patch") is attached to the opening of a micro-pipette with an orientation of the natural inside out or outside out, or a whole cell can be attached. Both sides of the membrane become accessible for experimental manipulation in biophysical studies, e.g. of solute concentrations at both sides of the membrane and of electrical gradients across the membrane. In plants it was important to isolate protoplasts from the cell wall to give the mouth of the micro-pipette access for measurements at the plasma membrane [66] and to use isolated vacuoles for measurements at the tonoplast [31,32,85].

- For groups of ion-channels and even for single channels the relationship between electrical current flowing across the membrane and the electrical potential across the membrane (voltage) can be characterized measuring the current/voltage relationships under the various experimental conditions chosen.
- Some channels can be passed in both directions. Often they are one way roads only in one direction, i.e. there is rectification outside-in or inside-out, which is an essential property revealed in patch-clamp measurements.
- Channels are not open continuously. There is gating where channels show a certain opening and closing probability. These dynamics are essential for functioning of channels. Patch-clamp measurements reveal these relationships.

Measuring trans-membrane electrical potentials with micro-electrodes provides information different from the types of information given by the patch clamp technique and is still needed, e.g. in studies of signaling and action potentials. However, the break through with the patch clamp technique and the revolution of molecular genetics which emerged somehow simultaneously joined in a union of biophysics and molecular biology fostering scientific advances now and in the future. Much information is currently obtained on the molecular structure of ion channels [30]. However, quantitative assessment of individual ion fluxes remains essential. This is accomplished by compartmentation analysis. Another particularly biophysical advancement for the quantification of ion fluxes was the development of ion-selective microelectrodes, e.g. for protons, potassium, sodium, ammonium, calcium, chloride and nitrate, applied to the measurement of solute ion uptake in a variety of plant organs under various conditions [73]. The molecular expression of the gene products of transporters must be verified at the functional level. Specific molecules and protein complexes of channels and transporters must be evaluated jointly by molecular biology and the biophysical electro-physiology. The current progress is documented in a detailed and remarkably comprehensive review by Hedrich [30]. This links membrane transporters to functioning of entire cells and whole-plant behavior [30,73].

Vision in Green Cells: The Eye of Chlamydomonas

We shall deal with the role of biophysics in photobiology below. However, at this stage it appears the best place to mention vision in green algal cells [33] with far reaching perspectives given by the union of molecular biology and patch-clamp biophysics.

During motion of cells of the unicellular flagellate alga *Chlamydomonas reinhardtii* rotating around their cell axis the orange red eye spot periodically darkens the photoreceptor which allows perception of the direction of light. The eye spot consists of a highly ordered array of carotenoid vesicles. It is not the actual eye of the alga. This is the photoreceptor in the plasma membrane which is a proton channel coupled to the chromophore rhodopsin, related to the retinal of our own eyes and named channel-rhodopsin. Its molecular structure is well characterized. Heterologous exprimation and patch clamp measurements have shown that light triggers a change of the channel from the closed to the opened state. This elicits three currents, a first current activity of the receptor channel and a faster and slower subsequent current which mediate coupling to the machinery of flagella movement. The photoreceptor is a unique light-activated H^+ ion channel functioning as a molecular light-switch [69]. It has received considerable interest in neurobiology because with heterologous exprimation in neurons it allows to use the light activation for studying complex switching between neurons [11].

A conclusion: Continuity of Plant-Membrane Biophysics from 1872 into the Age of Molecular Biology

Membrane transport was at the beginning of plant biophysics with a continuity in the development of electro-physiology by the first extracellular measurements (1872), the first trans-membrane measurements in giant algal cells (1930) and in cells of higher plants (1960) followed by the patch-clamp technique (1976), which now puts biophysics in the core of modern membrane research at the level of molecular dimensions.

Transport of Water: H_2O a Special Molecule

Basically the same rules apply to the transport of water molecules as to the solutes. Nevertheless as the major solvent water is a special case in life. Specific channels for the transport of water across membranes are the aquaporins [60,61], which molecular biology can isolate, characterize and engineer like the other channels discussed above. However, due to the essential role of water in all organisms and very conspicuously in plants water transport is a particular domain of biophysics [75].

At the cell level the Pfeffer-cell developed by Wilhelm Pfeffer (1877) is describing the distribution of water across a membrane:

$$\Delta\Psi = \Delta P - \sigma\Delta\pi \qquad (6)$$

Ψ is the water potential, as a key parameter in all water relations; $\Delta\Psi$ is the water potential gradient, ΔP is the gradient of hydrostatic pressure, which in plant cells is called turgor pressure. $\Delta\pi$ is the gradient of osmotic pressure given by concentrations of solutes. The cellular process of water transport driven by $\Delta\pi$ is called osmosis. The coefficient σ having values between 0 and 1 is characterizing membrane permeability. If the membrane is strictly semi-permeable, i.e. fully permeable for the water molecules and impermeable for solute molecules $\sigma = 1$. If the membrane also has non-restricted permeability for the solute molecules $\sigma = 0$. The water potential of a solution is given by the difference of the chemical activity of the water in the solution and the chemical activity of pure water μ^{0}_{H2O}:

$$\Psi_{solution} = \frac{\mu_{H2O} - \mu^{0}_{H2O}}{V_{H2O}} \qquad (7)$$

where V_{H2O} is the partial molar volume of water.

Physical methods are used for determining the parameters. Often π is measured cryoscopically. Ψ of leaves and shoot systems has been measured with the Scholander pressure chamber [82,99], where the organs are placed inside the chamber and the cut ends of shoots or petioles stick out through a tight seal. A gas pressure is applied to the chamber and the equilibrium pressure with the atmosphere where water is just starting to come out of the xylem vessels is taken to be equivalent to the water potentials of the organs. A more precise apparatus is the intracellular pressure probe [34,96], where an oil filled capillary is inserted into cells and the cellular turgor-pressure can be measured directly. As a xylem pressure probe the device can also be inserted into xylem vessels [65,106] or it can be attached to the cut ends of roots as a root pressure probe [65,95,108]. It allows measuring pressures and deduced variables including Ψ by using equation (6) under a wide range of experimental conditions.

At the whole plant level molecular labeling of aquaporins has been applied to show their involvement in water transport. Molecular transformation with the gene for luciferase [39] can be used as a reporter gene for the activity of the promoter of the aquaporin so that the intensity of bioluminescence elicited shows the spatiotemporal activity of the aquaporin [88]. Such molecular engineering allows conclusions about whole plant performance. Beyond that, however, the biophysical determination of water potential gradients is essential to understand the performance of plants in their environment. In addition to the water potential of solutions (equation 7) we need the water potential in the gas phase:

$$\Psi_{gas-phase} = \frac{\mu_{gas-phase} - \mu_{saturation}}{V_{H2O}} \tag{8}$$

where $\mu_{gas-phase}$ is the chemical activity of water in the atmosphere and $\mu_{saturation}$ the chemical activity at temperature dependent saturation. Physical meteorological tables provide values of the latter. The former is obtained from measurements of relative humidity in the atmosphere which are easy to perform (RH):

$$RH = \frac{\mu_{gas-phase}}{\mu_{saturation}} \cdot 100\ (\%) \tag{9}$$

In this way, water potential gradients can be assessed between solutions and the atmosphere, i.e. between tissues and organs within the plants and the environment. The $\Delta\Psi$ between the soil and leaves may be in the range of 1 MPa, while the $\Delta\Psi$ between leaves and the atmosphere can be 100 MPa [76]. Thus, there is an enormous driving force for water transport between the plants and the atmosphere, for the so called transpiration. It energizes the flow of water, i.e. the transpiration stream, in the conduits of the xylem. Adhesion of the dipole molecules of H_2O to the cell walls of the conducting elements and cohesion of the H_2O molecules between each other allow water columns to ascend under the tension of $\Delta\Psi$. This cohesion-tension theory appears to be the dominant mechanism quantitatively [43,98,106] although it is questioned by some authors [65,109]. Under some conditions osmotic processes along the path of the transpiration stream may also participate, but this is still debated [95,100]. A strong demand of further biophysical research is given to understand these principally and ecologically important problems.

Osmotic processes of root pressure and along the conduits of long distance transport in roots, stems and leaf-petioles also participate in refilling of conduits, i.e. in repair, after cavitation and embolism have occurred under stress of drought [15,21] and frost [63]. Structure-function relations governing the distribution of water in the whole plant along the soil-leaf continuum are the biophysical parameters of hydraulic architecture of plants [95], namely

- Specific stem conductivity K_s Ξ [kg s^{-1} m^{-1} MPa^{-1}],
- Cross section area per unit leaf area (Huber value) H_v [dimension less],
- Conductive stem per unit leaf area K_l Ξ [kg s^{-1} m^{-1} MPa^{-1}].

They are characterizing the environmental responses of plants and the eco-physiological regulation of plant water status.

PHOTOBIOLOGY

Light is electro-magnetic radiation. This is a domain of physics. Thus, evidently photobiology is biophysics. The two main fields of plant photobiology are photosynthesis and related processes and developmental biology. Photobiology of photosynthesis deals with quantitative aspects of radiation, i.e. the intensity of photosynthetically active photon flux density (PPFD) in the wavelength range of 400 to 750 nm. Photobiology of plant development deals with qualitative aspects of the regulatory effects of light of different wavelengths. Both are extraordinarily extensive areas of research and knowledge. It is far beyond the scope of the present assay to go to any details. However, in the endeavor of advocating perspectives of biophysics in plant biology we may not overlook them and need to allude to some general principles.

Photosynthesis

In photosynthesis chlorophyll molecules absorb photons, and electrons in the pigment molecules get excited to a first, second and third singlet state.

The latter two relax by the emission of heat. From the first singlet excited state the electrons flow via a chain of biochemical redox systems in the thylakoid membranes towards pyridine nucleotides (nicotineadenine-amide-dinucleotide, NAD, in some photosynthetically active bacteria, and nicotineadenine-amide-dinucleotide-phosphate, NADP, in other bacteria, cyanobacteria and eukaryotes) as acceptors. The reduced NAD(P)H drives the photochemical work of the reduction of CO_2 in assimilation and O_2 in photorespiration with the respective carboxylase and oxygenase functions of ribulose-bis-phosphate carboxylase/oxigenase (RubisCO) as well as reduction of NO_3^- and SO_4^{2-}. From the first excited singlet state electrons can also reach a triplet state from where energy transfer leads to the formation of reactive oxygen species (ROS) and the destructive photochemical work of photodamage.

From the first singlet excited state the chlorophyll can relax directly to the ground state by the emission of red light shifted to the far red region of the spectrum as compared to the 680 nm maximum of the excitation of chlorophyll *a* in photosystem II (PSII). This is called chlorophyll fluorescence. The intensity is weak. However, it can be readily measured by physical photometric techniques. It is an alternative to other pathways of energy dissipation and chlorophyll relaxation. As such it is a direct record of the condition and operation of PSII. Biophysical measurements of chlorophyll fluorescence have proven to be of enormous value in a huge range of studies in plant physiology and the physiological ecology of photosynthesis. Two equations are standing out, which may be given here to characterize the most important features of PSII in the dark adapted and light adapted states, respectively, of green photosynthetic cells and leaves:

$$\text{Potential quantum yield of PSII of dark adapted samples} = (F_m - F_0)/F_m \qquad (10)$$

$$\text{effective quantum yield of PSII of light adapted samples} = (F'_m - F)/F'_m \qquad (11)$$

where F_m and F'_m are the maximum fluorescence yields of the dark and light adapted samples, respectively, and F_0 and F_m give the respective ground fluorescence in weak measuring light [7,25,40,62,84]. From equation (11) the photosynthetic electron transport rate (ETR) can be derived as a quantitative measure:

$$ETR = 0.86 \times 0.5 \times [(F'_m - F)/F_m) \times PPFD] \qquad (12)$$

where the factor 0.86 indicates that on average 14 % of the incident radiation are reflected and the factor 0.5 accounts for distribution of radiation between the two photosystems, i.e. PSI in parallel to PSII.

Evidently studies of the primary reactions of light absorption and electron transport in photosynthesis are intrinsically biophysics. We should note in passing that design and construction of fluorometer equipment used requires a lot of physics. This also applies to porometers used for studying gas-exchange, i.e. of CO_2 and water vapor, in photosynthesis. This cannot be covered in the present assay but is of eminent importance for the links between plant biology and physics.

However, what we must remember is that the photosynthetic pigments and the redox molecules serving photosynthetic electron transport are imbedded in and interacting with proteins in the thylakoid membranes. A large number of different specific protein molecules are involved so that thylakoid proteomics is an important issue [24,107]. Once again in our survey, we encounter the need of close cooperation of biophysics and molecular biology.

Signaling and Development

As an example of light signaling I have already mentioned vision above. More generally light signaling for regulation of development is a vast field in plant biology. (In the brief survey here I follow Schopfer and Brennicke [83]). The biophysical steps are the absorption of light by various pigments. These are sensor pigments present only in low concentrations as compared to the mass pigments, such as the chlorophylls and other pigments in photosynthesis. The pigments are bound as chromophores to proteins forming the chromo-proteins effective in light signaling. There are two types of photosensors for near-ultraviolet and blue light in the wavelength-region of 340-520 nm, namely cryptochromes and phototropins, and phytochrome in red light, each with families of proteins. For example for phytochrome we can distinguish 5 forms of phytochrome A to E. The chromophores are pterine and a flavineadenine-dinucleotide in chryptochromes, a flavine-mononucleotide in phototropins and an open linear tetrapyrrole in phytochromes.

The activation of phytochromes occurs in a biophysical equilibrium of red light (R at 660 nm) and far-red light (FR at 730 nm). There is a red absorbing form, P_R, which after the absorption of R changes its conformation by a shift of the position of one of the terminal pyrrole rings and thus is transformed into the far-red absorbing form, P_{FR}. The latter returns to the P_R form after the absorption of FR:

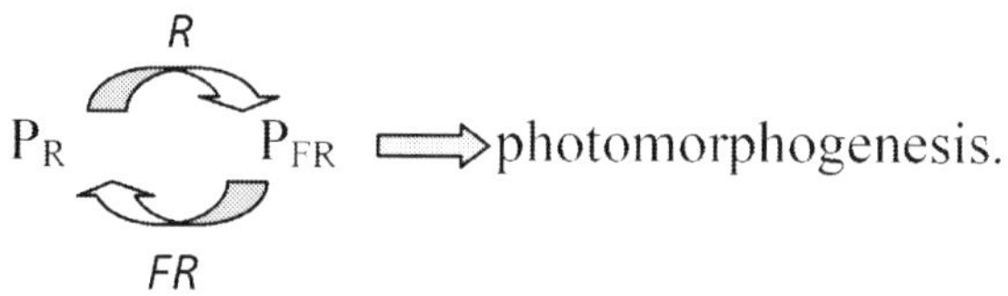

Under these light qualities the conversions between P_R and P_{FR} are strictly reversible. Quantitative radiation biophysics are involved in assessing phytochrome reactions at various light intensities as we distinguish three types, namely

- Very low fluence reactions,
- Low fluence reactions and
- High irradiance reactions.

The primary reactions of light absorption constitute the biophysical input pathways of signaling. Then there is a direct link to molecular biology. The pigments directly interact with the protein moieties of the chromo-proteins. For instance in phytochrome the change of the conformtion of the tetrapyrrole chain after absorption of R effects a conformation change of the phytochrome protein. This elicits a phosphorylation of the protein and its activation to the active P_{FR} form, which has the function of a kinase. In interaction with phosphorylases it can drive cascades of protein-phopshorylations in the cytosol. P_{FR} can also enter the nuclei to interact with regulators of gene activity. In this way downstream of the primary biophysical reaction of light absorption at the molecular level in cascades and in networks a plethora of events can be regulated which we summarize under the term of "photomorphogenesis".

Cryptochromes function in blue light responses of photomorphology and chronobiology. Phototropins are active in plant movements, e.g. phototropisms, and in the blue light regulation of stomatal aperture. The range of phytochrome regulated photomorphogeneses is etraordinarily broad. There are biophysical photo-electric membrane effects. Other examples are regulation of the synthesis of enzyme proteins, such as phenylalanine ammonia lyase for anthocyanine synthesis, nitrate and nitrite reductases, chalcone synthase and lipoxigenase. Further phytochrome-photomorphogeneses comprise differentiation of plastids, regulation of seed germination, etiolement/de-etiolement, nyctinastic movements of leaflets, photoperiodism of short day and long day plants.

Of enormous ecophysiological importance is the effect of shade under vegetation. With an absorption maximum of chlorophyll *a* in PSII at 680 nm photosynthesis filters out much of the red light necessary to stabilize the active P_{FR} form. In full sun light the R/FR-ratio is 1.2, but on a forest floor it may be reduced down to 0.5. The absorption spectra of P_R and P_{FR} are overlapping and it depends of the actual wavelength given how much P_{FR} is present for photomorphogeneses. There is a photo-equilibrium at any wavelength λ, i.e.

$$(\text{Photo-equilibrium})_\lambda = P_{FR\text{-}\lambda}/P_{\Sigma(PR+PFR)} \quad (13)$$

THE BIOLOGICAL CLOCK

The biological clock reveals itself by many endogenous rhythms in plants. These rhythms are circadian with period lengths close to 24 h when running free under constant environmental conditions [50]. Among several external control parameters of the clock we recognize blue and red light with input pathways via cryptochrome and phytochrome as receptors or "zeitnehmer". These are biophysical reactions (see above). However, current clock research is essentially molecular biology mostly using *Arabidopsis thaliana*. Among the master genes of the clock there are the morning genes *Circadian Clock Associated (CCA1)* and *LATE ELONGATED HYPOCOTYL (LHY)* and the evening genes *TIMING OF CHLOROPHYLL a/b BINDING (TOC1)* in the evening. Downstream there are additional morning and evening elements functioning as transcription factors [28,38,64,70] and a vast number of clock controlled genes (CCGs) putting a plethora of plant functions under the regime of the clock [23].

An important circadian rhythm among plants is the metabolic rhythm of crassulacean acid metabolism (CAM), and here we have identified a biophysical oscillator operating in close interdependence with the molecular oscillators. CAM is an adaptation to problems of water supply because CO_2 is fixed by phosphor *enol* pyruvate carboxylase (PEPC) in the dark period, when evaporative water demand is low, leading to the formation of malic acid which is stored in the central cell vacuole over night. The malic acid is remobilized during the light period, decarboxylated behind closed stomata producing CO_2 for photosynthetic assimilation. The endogenous biological clock is behind it, since the cycle is also running under constant environmental conditions in continuous light or darkness.

PEPC is phosphorylated in its active state and the PEPC-kinase is under circadian control [17,29]. Therefore, for some while it was assumed that the endogenous CAM-rhythm is specifically a molecular process under the genetic control of PEPC-kinase expression. By contrast, we had proposed a biophysical oscillator. It is based on a tension/relaxation mechanism of the tonoplast membrane given by the osmotic consequences of bulk vacuolar accumulation and remobilization, respectively, of malic acid. The biophysical hysteresis switch or beat oscillator is a switch between states of order of the tonoplast with (i) low order/high fluidity/high permeability for malic acid at high malic acid filling state and higher turgor pressure facilitating malic acid remobilization, and (ii) high order/low fluidity/low permeability for malic acid at malic acid empty state and lower turgor pressure facilitating malic acid accumulation, respectively [48,55]. Thus, the tonoplast is functioning as a biophysical master switch in the circadian rhythm of CAM [49].

Later studies showed that the molecular oscillator of the genetic control of PEPC-kinase and the biophysical oscillator of tonoplast state of membrane order are coupled. The gene expression of PEPC-kinase was found to be under the metabolic control of malate in the cytoplasm [10,74]. This constitutes a direct link to the biophysical malic acid influx/efflux oscillator affecting cytoplasmic malate concentrations. The endogenous rhythm of CAM is an excellent example of the integration of both biophysical and molecular mechanisms in complex physiological functions.

WHOLE PLANT INTEGRATION: SELF-ORGANIZATION AND EMERGENCE

The building blocks of plants are the organs roots, shoots and leaves. Building blocks of the organs are various tissues. We may consider these as modules. The higher organized systems, such as the organs or the entire plants are more than the sum of their parts. There is integration with self-organization from the parts. Plants are integral organisms. In these individuals properties and capacities are expressed which go far beyond the modularity of the parts. This result of self-organization is also named emergence [51]. The unitary integration of modules in the integral units of plants requires perception and transduction of signals and communication, i.e. sharing of information [52,87]. Basically there are three types of signals, chemical signals, electrical signals, hydraulic signals. The latter two are biophysical. Understanding whole plant integration is an eminently biophysical task.

Chemical signals are the primary messengers of the classic groups of phytohormones and secondary messengers, where currently increasing numbers of regulatory compounds are identified as the understanding of the complexity of networks of physiological and biochemical functions and their integration advances.

Electrical signals are action potentials (AP), variation or slow wave potentials (VP) and system potentials (SP) with rates of propagation ranging from just below 1 mm s^{-1} to 50 mm s^{-1} or even higher [52]. The systemic function of long distance electrical signaling in plants requires their transmission which occurs mainly in the long distance transport bundles and particularly in the phloem [26,27,52,86].

Hydraulic signals are pressure changes and hydraulic waves [26,59,97] with the movement of water columns in the xylem [18]. In the cells membranes, particularly the plasma membrane and the tonoplast, are the sites of sensing the turgor pressure (P) which is determined by gradients of water potential (Ψ) and osmotic potential (π) (equation 6). Ψ cannot be perceived by the cells directly but indirectly via P and π. There are three ways in which in shoots and leaves hydraulic signals of the roots can be decoded: (i) via osmosensing π, (ii) via sensing P, and (iii) by sensing mechanical forces at the cell wall or even in membranes via mechano-sensitive stretch responsive ion channels and ion pumps [52,87,91,92].

In signal transduction chains and networks the three types of signals, the chemical and the two biophysical ones can be translated into each other, there is a cross talk between physical and chemical signals. A couple of examples may show *"connections between physical, (bio) chemical and molecular signals in the plant"* [87].

Gravitropic orientation of the plant is a developmental process governed by electric and chemical signaling. Gravitropic root bending is elicited by sensing the position of the root in relation to the field of gravity in the root caps within special cells, the statocytes, by starch grains lying on a cushion of endoplasmic reticulum and functioning as statoliths. An early event observed in gravitropic reactions of roots is the so-called geo-electric effect first described by Brauner and Bünning [12] and later confirmed in detailed studies by Behrens et al. [6] and Stenz and Weisenseel [93,94], where the membrane potentials in root cap cells at the physically upper face of a root coming into a horizontal position in relation to gravity become more negative (-131 mV) than at the lower face (-93 mV) [6]. The sensing is in the tip of the root but the bending occurs by asymmetric growth further up in the elongation zone. This requires signaling which involves asymmetric distribution of the phytohormone indole-acetic acid (IAA) coming into the root from above and being transported back and of a second phytohormone cytokinine produced in the root tip [2].

An important mechanism in whole plant integration is root-shoot signaling of water relations where the roots can inform about water problems for early stomatal regulation in the leaves even if these have not yet suffered from reduced supply. Hydraulic and chemical signals can translate into electrical signals. Hydraulic surges and changes of turgor pressure are translated into electrical signals of APs, VPs and SPs [20,52]. Hydraulic signals interact with chemical signaling via the phytohormone abscisic acid (ABA) and another phytohormone interacting with root-shoot electrical signaling is cytokinin [86].

BIONICS

Bionics links biology with technology. The term biomimetics is also used, but bionics is much more than simple mimetic imitation or copying of nature. Werner Nachtigall is distinguishing the two steps of technical biology and bionics [67]. The first one is the step from recognition of structure-function relations in the living world by the biologist with extraction of general principles leading to modeling by analog abstraction from the original biological data. It is a field in its own right but not yet bionics. Bionics then is the step from abstraction towards appropriate technical realization by the constructing engineer. There is self-acceleration in the process. Biological recognition leading to technical applications and new technical capabilities via bionics feeds back via technical biology to biological understanding. Engineering intrinsically uses physical approaches.

Structural and functional biophysics of plants giving ideas for technical developments include construction of wheels, features of wings, building principles of the hollow stems of grasses, statics of buttress roots, and effects of surfaces [8,13,56,68,89].

The wheel has probably originated from cut sections of the stems of trees. Recently super-light wheels have been developed after the three-dimensional structure of spherical diatoms. One-wing flying machines have been designed after the giant up to 15 cm wide wind dispersed winged seeds of the Cucurbitaceae *Zanonia macrocarpa*. The technical grass shoot has been constructed after the tall grass *Arundo donax* as well as the horsetail *Equisetum*, and his light building-principle combined with mechanical strength has found many technical applications as tubes for special purposes and in constructions of many areas of engineering [89,90,102]. Buttress roots stabilize large trees and this construction principle has been analyzed for the application in reducing the notch stress in buildings and other technical constructions [58]. Studies of the sculpturing of surfaces of plants, especially of some leaves, have led to bionic surface technology [3,4,5]. Due to being covered with epicuticular wax crystals of about 110 nm diameter surfaces of leaves become super-hydrophobic, i.e. anti-adhesive for droplets of water. They are self-cleaning because the water droplets running off take mud and dirt particles along with them. The most widely known example is the leaves of the lotus plant *Nelumbo nucifera* which is a holy plant and symbol of purity in Buddhism and Hinduism. It has led to the bionic development of surface paints for self-cleaning facades of buildings, for automobiles and other surfaces technically known under the trademark Lotus-Effect[R]. Another example of super-hydrophobic plant surfaces are the leaves of the floating water fern *Salvinia*. The highly complicated structured surfaces have super-hydrophobic elastic hairs between which an air layer is kept under water. Ships covered with such air-retaining surfaces would benefit from the Salvinia-Effect by significantly reduced friction when gliding through water which would help to reduce fuel consumption.

On the molecular level it is attempted to realize the dream of using the principle of photosynthetic electron transport in thylakoid membranes for technical provision of energy by a biomimetic kind of photo-voltaics. In the so-called Graetzel-cell two glass electrodes, anode and cathode, are used, where the anode contains pigment molecules triggering a light-driven flow of electrons, i.e. current [77]. Initially these were actually chlorophyll molecules; later they were replaced by synthetic pigments. It is a promising biophysical approach deserving more study and technical development [105].

LIFE IN SPACE: AN EXTREME BIOPHYSICAL PERSPECTIVE

Exploration of space is the realm of physics, astrophysics. Speculations about life elsewhere in the universe are nourished by the statistical argument that the increasing discovery of huge numbers of earth-like planets in the universe makes it likely that life exists elsewhere although distances are so immense that we shall never have a chance to learn about that. The question is immensely biophysical: first, because of the astrophysical description of space; second, as life elsewhere would also need primary producers feeding it and all the physical processes of energy transduction involved, and would be subject to thermodynamics of dynamic steady state "equilibria".

OUTLOOK: QUO VADIS PLANT-BIOPHYSICS IN THE MOLECULAR AGE?

This brief list of some hot-spots of biophysics in plant biology and tour of the horizon shows that biophysics is indispensible for putting together the puzzles of modules obtained in trans-sub-disciplines of plant biology to understand emergence of higher integrated systems. It is necessary for comprehending plant life and assessing the important services plants make to the function of other living systems, such as biomes and ecosystems including agro-and forest ecosystems, sustaining the existence also of man on our globe. Biophysics is the basis of many innovations, such as bio-mimetic technologies. Plant-biophysics is not fading away in the molecular age. By the very contrary, all the hot spots show that it must merge with molecular biology to really foster progress of basic understanding and sustainable management of plant life (Figure 1). In quite a number of cases as shown in this article such union is already effective and proves to have considerable potential for managing the future. Sustained and increased practice of biophysics with teaching and research deserves strong encouragement.

REFERENCES

1. Abbaspour N, Kaiser B, Tyerman S (2013) Chloride transport and compartmentation within main and lateral roots of two grapevine rootstocks differing in salt tolerance. *Trees* 27: 1317–1325.
2. Aloni R, Langhans M, Aloni E, et al. (2004) Role of cytokinin in the regulation of root gravitropism. *Planta* 220: 177–182.
3. Barthlott W, Neinhuis C (1997) Purity of the sacred lotus for escape from contamination in biological surfaces. *Planta* 202: 1–8.
4. Barthlott W, Mail M, Neinhuis C (2016) Superhydrophobic hierarchically structured surfaces in biology: evolution, structural principles and biomimetic applications. *Philos T Roy Soc A*. DOI: 10.1098/rsta.2016.0191.
5. Barthlott W, Schimmel T, Wiersch S, et al. (2010) The Salvinia paradox: Superhydrophobic surfaces with hydrophilic pins for air-retention under water. *Adv Eng Mater* 22: 1–4.
6. Behrens HM, Gradmann D, Sievers A (1985) Membrane potential responses following gravistimulation in roots of *Lepidium sativum* L. *Planta*163: 463–472.
7. Bilger W, Schreiber U, Bock M (1995) Determination of the quantum efficiency of photosystem II and the non-photochemical quenching of chlorophyll fluorescence in the field. *Oecologia* 102: 425–432.
8. Blüchel KG, Malik F (2006) Faszination Bionik: Die Intelligenz der Schöpfung. Mcb Verlag, St. Gallen
9. Böhm J, Scherzer S, Krol E, et al. (2016) The Venus flytrap *Dionea muscipula* counts prey-induced action potentials to induce sodium uptake. *Curr Biol* 26: 286–295.
10. Borland AM, Hartwell J, Jenkins GI, et al. (1999) Metabolite control overrides circadian regulation in phosphoenolpyruvate carboxylase kinase and CO_2 fixation in Crassulacean acid metabolism. *Plant Physiol* 121: 889–896.
11. Boyden ES, Zhang F, Bamberg E, et al. (2005) Millisecond-

timescale, genetically targeted optical control of neuronal activity. *Nat Neurosci* 8: 1263–1268.

12. Brauner L, Bünning E (1930) Geoelektrischer Effekt und Elektrotropismus. *Berichte der Deutschen Botanischen Gesellschaft* 48: 470–476.
13. Brickwede F, Erb R, Lefèvre J, et al. (2007) Bionik und Nachhaltigkeit-Lernen von der Natur. Erich Schmidt Verlag, Eds., Berlin.
14. Britto DT, Ruth TJ, Lapis S, et al. (2004) Cellular and whole-plant chloride dynamics in barley: insights into chloride-nitrogen interactions and salinity responses. *Planta* 218: 615–622.
15. Bucci SJ, Scholz FG, Goldstein G, et al. (2003) Dynamic changes in hydraulic conductivity in petioles of two savanna tree species: factors and mechanisms contributing to the refilling of embolized vessels. *Plant Cell Environ* 26: 1633–1645.
16. Burdon-Sanderson J (1872) Note on the electrical phenomena which accompany stimulation of the leaf of *Dionaea muscipula* Ellis. *Philos Proc Roy Soc* 21: 495–496.
17. Carter PJ, Nimmo HG, Fewson CA, et al. (1991) Circadian rhythms in the activity of a plant protein kinase. *Embo J* 10: 2063–2068.
18. Cermak J, Matyssek R, Kucera J (1993) Rapid response of large, drought –stressed beech trees to irrigation. *Tree Physiol* 12: 281–290.
19. Chen ZH, Hills A, Bätz U, et al. (2012) Systems dynamic modeling of the stomatal guard cell predicts emergent behaviors in transport, signaling, and volume control. *Plant Physiol* 159: 1235–1251.
20. Comstock JP (2002) Hydraulic and chemical signaling in the control of stomatal conductance and transpiration. *J Exp Bot* 53: 195–200.
21. Domec JC, Scholz FG, Bucci SJ, et al. (2006) Diurnal and seasonal variation in root xylem embolism in neotropical savanna woody species: impact on stomatal control of plant water status. *Plant Cell Environ* 29: 26–35.
22. Etherton B, Higinbotham N (1960) Transmembrane potential measurements of cells of higher plants as related to salt uptake. *Science* 131: 409–410.

23. Farré EM (2012) The regulation of plant growth by the circadian clock. *Plant Biology* 14: 401–410.
24. Friso G, Giacomelli L, Ytterberg AJ, et al. (2004) In-depth analysis of the thylakoid membrane proteome of *Arabidopsis thaliana* chloroplasts: new proteins, new functions, and a plastid proteome database. *Plant Cell* 16: 478–499.
25. Genty B, Briantais JM, Baker NR (1989) The relationship between quantum yield of photosynthetic electron transport and quenching of chlorophyll fluorescence. *BBA* 990: 87–92.
26. Grams TEE, Koziolek C, Lautner S, et al. (2007) Distinct roles of electric and hydraulic signals on the reaction of leaf gas exchange upon re-irrigation in *Zea mays* L. *Plant Cell Environ* 30, 79–84.
27. Grams TEE, Lautner S, Felle HH, et al. (2009) Heat-induced electrical signals affect cytoplasmic and apoplastic pH as well as photosynthesis during propagation through the maize leaf. *Plant Cell Environ* 32: 319–326.
28. Harmer SL, Kay SA (2005) Positive and negative factors confer phase-specific circadian regulation of transcription in *Arabidopsis. Plant Cell*17: 1926–1940.
29. Hartwell J, Smith LH, Wilkins MB, et al. (1996) Higher plant phosphoenolpyruvate carboxylase kinase is regulated by the level of translatable mRNA in response to light or a circadian rhythm. *Plant J* 10: 101–108.
30. Hedrich R (2012) Ion channels in plants. *Physiol Rev* 92: 1777–1811.
31. Hedrich R, Schroeder JI, Fernandez JM (1987) Patch-clamp studies on higher plant cells: a perspective. *Trends Biochem Sci* 12: 49–52.
32. Hedrich R, Barbier-Brygoo H, Felle H, et al. (1988) General mechanisms for solute transport across the tonoplast of plant vacuoles: a patch-clamp survey of ion channels and proton pumps. *Bot Acta* 101: 7–13.
33. Hegemann P (1997) Vision in microalgae. *Planta* 203: 265–274.
34. Hüsken D, Steudle E, Zimmermann U (1978) Pressure probe technique for measuring water relations of cells of higher plants. *Plant Physiol*61: 158–163.

35. Hütt M-Th (2013) A network view on patterns of gene expression and metabolic activity. *Nova Acta Leopoldina* NF 114/391: 183–199.
36. Hütt M-Th, Lüttge U (2002) Nonlinear dynamics as a tool for data analysis and modelling in plant physiology. *Plant Biology* 4: 281–297.
37. Hütt M-Th, Lüttge U (2004) Network dynamics in plant biology: Current progress in historical perspective. *Prog Bot* 66: 277–310.
38. Kikis EA, Khanna R, Quail PH (2005) ELF4 is a phytochrome-regulated component of a negative-feedback loop involving the central oscillator components CCA1 and LHY. *Plant J* 44: 300–313.
39. Koo J, Kim Y, Kim J, et al.(2007). A GUS/luciferase fusion reporter for plant gene trapping and for assay of promoter activity with luciferin-dependent control of the reporter protein stability. *Plant Cell Physiol* 48: 1121–1131.
40. Koten O van, Snel JFH (1990) The use of chlorophyll fluorescence nomenclature in plant stress physiology. *Photosynth Res* 25: 147–150.
41. Kronzucker HJ, Siddiqi MY, Glass ADM, et al. (1999) Nitrate-ammonium synergism in rice. A subcellular flux analysis. *Plant Physiol* 119: 1041–1045.
42. Laughlin RB (2005) A Different Universe—Reinventing Physics from the Bottom Down. *Basic Books,* New York.
43. Lösch R (1998) Plant water relations. *Prog Bot* 60: 193–233.
44. Lundegardh H (1950) The translocation of salts and water through wheat roots. *Plant Physiol* 2: 103–151.
45. Lundegårdh H (1955) Mechanisms of absorption, transport, accumulation, and secretion of ions. *Annu Rev Physiol* 6: 1–24.
46. Lundegårdh H, Burström H (1933) Untersuchungen über die Salzaufnahme der Pflanzen. III. Quantitative Beziehungen zwischen Atmung und Anionenaufnahme. *Biochemische Z* 261: 235–251.
47. Lundeghårdh H, Burström H (1935) Untersuchungen über die Atmungsvorgänge in Pflanzenwurzeln. *Biochemische Z* 277: 223–249.
48. Lüttge U (1986) Nocturnal water storage in plants having

crassulacean acid metabolism. *Planta* 168: 287–289.

49. Lüttge U (2000) The tonoplast functioning as the master switch for circadian regulation of crassulacean acid metabolism. *Planta* 211: 761–769.
50. Lüttge U (2003) Circadian rhythmicity: Is the "biological clock" hardware or software. *Prog Bot* 64: 277–319.
51. Lüttge U (2012) Whole-plant physiology: Synergistic emergence rather than modularity. *Prog Bot* 74: 165–190.
52. Lüttge U (2013) Modularity and emergence: biology's challenge in understanding life. *Plant Biology* 14: 865–871.
53. Lüttge U (2016) Transport processes—The key integrators in plant biology. *Prog Bot* 77: 3–65.
54. Lüttge U, Hütt MT (2009) Talking Patterns: Communication of organisms at different levels of organization—an alternative view of systems biology. *Nova Acta Leopoldina* NF 96/357: 161–174.
55. Lüttge U, Kluge M, Ball E (1975) Effects of osmotic gradients on vacuolar malic acid storage. A basic principle in oscillatory behavior of crassulacean acid metabolism. *Plant Physiol* 56: 613–616.
56. Lüttge U, Kluge M, Thiel G (2010) Botanik. Die umfassende Biologie der Pflanzen. Wiley-VCH.
57. MacRobbie EAC (1965) The nature of the coupling between light energy and active ion transport in *Nitella translucens*. *BBA* 94: 64–73.
58. Mattheck C (2006) Verborgene Gestaltgesetze in der Natur. Forschungszentrum Karlsruhe.
59. Matyssek R, Maruyama S, Boyer JS (1991) Growth-induced water potentials may mobilize internal water for growth. *Plant Cell Environ* 14: 917–923.
60. Maurel C (1997) Aquaporins and water permeability of plant membranes. *Annu Rev Plant Physiol Plant Mol Biol* 48: 399–429.
61. Maurel C, Santoni V, Luu DT, et al. (2009) The cellular dynamics of plant aquaporin expression and functions. *Curr Opin Plant Biol* 12: 690–698.
62. Maxwell K, Johnson GN (2000) Chlorophyll fluorescence—a practical guide. *J Exp Bot* 51: 659–668

63. Mayr S, Améglio T (2016) Freezing stress in tree xylem. *Prog Bot* 77: 381–414.

64. McClung CR (2006) Plant circadian rhythms. *Plant Cell* 18: 792–803.

65. Melcher PJ, Meinzer FC, Yount DE, et al. (1998) Comparative measurements of xylem pressure in transpiring and non-transpiring leaves by means of the pressure chamber and the xylem pressure probe. *J Exp Bot* 49: 1757–1760.

66. Moran N, Ehrenstein G, Iwasa K, et al. (1984) Ion channels in plasmalemma of wheat protoplasts. *Science* 226: 835–838.

67. Nachtigall W (2010): Bionik als Wissenschaft: Erkennen—Abstrahieren—Umsetzen. Springer Verlag, Heidelberg.

68. Nachtigall W, Blüchel KG (2000) Das große Buch der Bionik. Deutsche Verlagsanstalt, Stuttgart.

69. Nagel G, Ollig D, Fuhrmann M, et al. (2002) Channelrhodopsin-1: A light-gated proton channel in green algae. *Science* 296: 2395–2398.

70. Nakamichi N (2011) Molecular mechanisms underlying the *Arabidopsis* circadian clock. *Plant Cell Physiol* 52: 1709–1718.

71. Neher E, Sakmann B (1976) Single channel currents recorded from membrane of denervated frog muscle fibers. *Nature* 260: 799–802.

72. Neher E, Sakmann B, Steinbach JH (1978) The extracellular patch clamp: A method for resolving currents through individual open channels in biological membranes. *Pflüg Arch* 375: 219–228.

73. Newman IA (2001) Ion transport in roots: measurement of fluxes using ion-selective microelectrodes to characterize transporter function. *Plant Cell Environ* 24: 1–14.

74. Nimmo HG (2000) The regulation of phospho*enol*pyruvate carboxylase in CAM plants. *Trends Plant Sci* 5: 75–80.

75. Nobel PS (2009) Physicochemical and environmental plant physiology, 4Eds. Amsterdam: Academic Press.

76. Nobel PS, Jordan PW (1983) Transpiration stream of desert species: resistances and capacitances for a C_3, a C_4 and a CAM plant. *J Exp Bot* 34: 1379–1391.

77. O'Regan B, Grätzel M (1991) A low-cost, high-efficiency solar cell based on dye-sensitized colloidal TiO_2 films. *Nature* 353: 737–740.

78. Pitman MG (1963) The determination of the salt relations of the cytoplasmic phase in cells of beetroot tissue. *Aust J of Biol Sci* 1: 647–668.

79. Raven JA (1967) Light stimulation of active transport in *Hydrodictyon africanum*. *J Gen Physiol* 50: 1627–1640.

80. Raven JA (1976) Transport in algal cells. *Encyclopedia Plant Physiology New Series* 2A, Springer Berlin Heidelberg, 129–188.

81. Sakmann B, Neher E (1984) Patch clamp techniques for studying ionic channels in excitable membranes. *Annu Rev Physiol* 46: 455–472.

82. Scholander P, Bradstreet E, Hemmingsen E, et al (1965) Sap pressure in vascular plants: Negative hydrostatic pressure can be measured in plants. *Science* 148: 339–346.

83. Schopfer P, Brennicke A (2010) *Pflanzenphysiologie.* 7Eds. Spektrum, Heidelberg.

84. Schreiber U, Bilger W (1993) Progress in chlorophyll fluorescence research: major developments during the past years in retrospect. *Prog Bot* 54: 151–173.

85. Schroeder JI, Hedrich R, Fernandez JM (1984) Potassium-selective single channels in guard cell protoplasts of *Vicia faba*. *Nature* 312: 361–362.

86. Shabala S, Pang J, Zhou M, et al. (2009) Electrical signalling and cytokinins mediate effects of light and root cutting on ion uptake in intact plants. *Plant Cell Environ* 32: 194–207.

87. Shabala S, White RC, Djordjevic MA, et al. (2016) Root-to-shoot signalling: integration of diverse molecules, pathways and functions. *Funct Plant Biol* 43: 87–104.

88. Siefritz F, Otto B, Bienert G, et al. (2004), The plasma membrane aquaporin NtAQP1 is a key component of the leaf unfolding mechanism in tobacco. *Plant J* 37: 147–155.

89. Speck T, Speck O, Neinhuis C, et al. (2011) Was die Technik von Pflanzen lernen kann–Bionik in botanischen Gärten. Verband Botanischer Gärten, Freiburg, Dresden und Bayreuth.

90. Speck T, Neinhuis C (2004) Bionik, Biomimetik–Ein

interdisziplinäres Forschungsgebiet mit Zukunftspotential. *Naturwissdenschaftliche Rundsch* 57: 177–191.

91. Stancović B, Zawadzki T, Davies E (1997) Characterization of the variation potential in sunflower. *Plant Physiol* 115: 1083–1088.
92. Stancović B, Witters DL, Zawadzki T, et al. (1998) Action potentials and variation potentials in sunflower: an analysis of their relationships and distinguishing characteristics. *Physiol Plantarum* 103: 51–58.
93. Stenz HG, Weisenseel MH (1991) DC-electric field affects the growth direction and statocyte polarity of root tips (*Lepidium sativum*). *J Plant Physiol* 138: 335–344.
94. Stenz HG, Weisenseel MH (1993) Electrotropism of maize (*Zea mays* L.) roots. Facts and artifacts. *Plant Physiol* 101: 1107–1111.
95. Steudle E (2011) Hydraulic architecture of vascular plants. In: Lüttge U, Beck E, Bartels D, *Plant desiccation tolerance*, Springer-Heidelberg, 185–207.
96. Steudle E, Smith JAC, Lüttge U (1980) Water relation parameters of individual mesophyll cells of the Crassulacean Acid Metabolism plant *Kalanchoë daigremontiana*. Plant Physiol 66: 1155–1163.
97. Tang AC, Boyer JS (2003) Root pressurization affects growth-induced water potentials and growth in dehydrated maize plants. *J Exp Bot* 54: 2479–2488.
98. Tyree MT (1997) The cohesion-tension theory of sap ascent: current controversies. *J Exp Bot* 48: 1753–1765.
99. Tyree MT, Hammel HT (1972) The measurement of the turgor pressure and the water relations of plants by the pressure –bomb technique. *J Exp Bot* 23: 267–282.
100. Tyree MT, Zimmermann MH (2002) Xylem structure and the ascent of sap. Springer, Berlin.
101. Umrath K (1930) Untersuchungen über Plasma und Plasmaströmung an Characeen. IV. Potentialmessungen an *Nitella mucronata* mit besonderer Berücksichtigung der Erregungsleitung. *Protoplasma* 9: 576–597.
102. Vogel S (2000) Von Grashalmen und Hochhäusern–Mechanische Schöpfungen in Natur und Technik. Wiley-VCH.

103. Walker NA (1976) Membrane transport: Theoretical background. *Encyclopedia of Plant Physiology New Series* 2A, 36–52.

104. Walker NA, Pitman MG (1976) Measurement of fluxes across membranes. *Encyclopedia of Plant Physiology New Series* 2A, 93–126.

105. Wei J, Hongzhong L, Lei Y, et al. (2016) Enhanced photoelectric properties in dye-sensitized solar cells using TiO_2 pyramid arrays. *J Phys Chem* 120: 9678–9684.

106. Wei MC, Tyree MT, Steudle E (1999) Direct measurement of xylem pressure in leaves of intact maize plants. A test of the cohesion-tension theory taking hydraulic architecture into consideration. *Plant Physiol* 121: 1191–1205.

107. Whitelegge JP (2003) Thylakoid proteomics. *Photosynth Res* 78: 265–277.

108. Zhu GL, Steudle E (1991) Water transport across maize roots: simultaneous measurements of flows at the cell and root level by double pressure probe technique. *Plant Physiol* 95: 305–315.

109. Zimmermann U, Meinzer FC, Benkert R, et al. (1994) Xylem water transport: is the available evidence consistent with the cohesion theory. *Plant Cell Environ* 17:1169–1181.

CHAPTER 10

PLANTS AS MODEL IN BIOMIMETICS AND BIOROBOTICS: NEW PERSPECTIVES

Barbara Mazzolai, Lucia Beccai, andVirgilio Mattoli

Center for Micro-BioRobotics, Istituto Italiano di Tecnologia, Pontedera, Italy

ABSTRACT

Especially in robotics, rarely plants have been considered as a model of inspiration for designing and developing new technology. This is probably due to their radically different operational principles compared to animals and the difficulty to study their movements and features. Owing to the sessile nature of their lifestyle, plants have evolved the capability to respond to a wide range of signals and efficiently adapt to changing environmental conditions. Plants in fact are able to show considerable *plasticity* in their morphology and physiology in response to variability within their environment. This results in movements that are characterized by energy efficiency and high density. Plant materials are optimized to reduce energy consumption during motion and these capabilities offer a plethora of solutions in the artificial world, exploiting approaches that are muscle-free and thus not necessarily animal-like. Plant roots then are excellent natural diggers, and their characteristics such as adaptive growth, low energy consumption movements, and the capability of penetrating soil at any angle are interesting from an engineering perspective. A few examples are described to lay the perspectives of plants in the artificial world.

Keywords: Plant inspired solutions, soft robotics, biomimetics, growing robots, biorobotics

PLANTS AS MODELS IN THE ARTIFICIAL WORLD

The study and the extraction of biological key principles, and their translation in design guidelines for a new generation of robots and technological solutions, have been widely adopted in biomimetics and bionics (Gao et al., 2005; Fratzl and Barth, 2009; Zang et al., 2013). Biologically inspired approaches have been traditionally widely adopted in robotics as well (Ayers and Witting, 2007; Ijspeert et al., 2007; Cutkosky and Kim, 2009; Cianchetti et al., 2011). Robots that implement solutions inspired by nature show capabilities that permit adaptive, flexible interactions with unpredictable environments (Pfeifer et al., 2012; Kim et al., 2013). In the goal to mimic living beings, the different components of an artificial system have to be designed from models of the reference biological systems, in an integrated way, making explicit the general design principles underlying the embodiment. The next generation of robots will be "soft," because this better allows the interaction with environment, mediated by body, as in natural systems (Tramacere et al., 2012). Advances in "soft technology" will lead to a quantum leap in intelligent robotics. "Soft Robotics" will express at different levels: in sensing (e.g., soft skin to touch, deformable tissue, materials); in movement (e.g., elastic, compliant materials for muscles, tendons; variable compliant actuators, etc.); in interaction (e.g., soft movements, social and cognitive skills); in emotions (e.g., facial expression, body posture, etc.) (Pfeifer and Bongard, 2007).

Traditionally, the most studied models in biorobotics and soft robotics are animals, for a plethora of characteristics or for specific applications (Pfeifer and Bongard, 2007; Mazzolai et al., 2012; Pfeifer et al., 2012; Hawkes et al., 2013; Ma et al., 2013). Rarely, plants have been considered a source of inspiration for robotics or, more in general, for innovative engineering solutions. This is probably due to their radically different operational principles compared to animals and the difficulty to study their movements and features. Nevertheless, owing to the sessile nature of their lifestyle, plants have evolved the capability to respond to a wide range of signals and efficiently adapt to changing environmental conditions. Although plants lack muscles, they generate several different kinds of movements that span from hours or days to milliseconds (Martone et al., 2010; Forterre, 2013). The question of how plants can achieve such a wide range of non-muscular movements has attracted the interest of many scientists, since the pioneering work of Darwin (Darwin, 1880; Hart, 1990; Dumais and Forterre, 2012). An understanding of these non-muscular movements could hold the potential for developments in applied sciences and engineering, especially for the fabrication of novel biomimetic actuation strategies and bioinspired devices (Burgert and Fratzl, 2009).

Plants or plant parts, such as roots or leaves, offer countless cues for making innovation in technology. Some of the main principles studied in the plant roots and, more in general, in plants, in order to be applied to the development of new technologies can be summarized as follows:

- *Capacity of growth and movement* in response to external stimuli with high plasticity and morphological adaptation to the environment. This implies a character that can be: nastic (movement that is independent of the spatial direction of a stimulus) or tropic (the response of plant is influenced by the direction of a stimulus); active (live plant cells activate and control the response by moving ions and by changing the permeability of membranes based on potential actions) or passive (movements that are based on dead tissue that is suitable to undergo predetermined modifications upon changes in environmental conditions); reversible or irreversible. This implies the development of new actuation solutions for steering or elongating robotic parts;
- *Osmotic actuation*, used for triggering fast movements or driving slow movements in plants. A new generation of actuators, characterized by low power consumption and high

energy density can be derived by this fundamental plant property;

- *Sensory capabilities* with respect to a wide amount of different physical and chemical quantities in the environment (especially soil). Innovative sensing systems that incorporate at the micro-scale several transduction mechanisms for detecting several parameters, which are intelligently "interpreted" for efficient behavior, are promising technological solutions;
- *Growth from the tip* of the root by adding cells and production of lateral hairs, which reduces friction and pressure needed to penetrate the soil. New bioinspired robotic artifacts that grow by adding new materials are emerging;
- *Emergent behavior* given by coordination of the root apices of the whole organism toward optimal targets. This research can lead to develop novel methods of collective decision making in decentralized structures with local computation and simple communication.

In the following, we highlight a few examples of solutions inspired by plants, including materials, actuations, and plant root-like robots.

NEW MATERIALS AND ACTUATORS INSPIRED BY STRUCTURES AND MOTIONS IN PLANTS

Nowadays one of the major boosts in material science is given by inspiration from nature. The inspiration to natural solution allows the development of new materials with enhanced structural properties, applicable in several fields (including consumer products, automotive, and architecture) and with sensing and actuation capability.

The interesting aspect of natural materials is related to the numerous functionalities that they can exploit (e.g., stability, stiffness, toughness, self-healing, etc.), even if they are based on a limited number of basic components (i.e., cellulose, hemicellulose, lignin, and pectin). This variety of properties is mainly due to the hierarchical organization of such components. Extraction of the principles underlying these properties is one of the key approaches of bioinspired materials.

Of particular interest are those materials and principles that are inspired by plant world, since they are still largely unexploited (except for few well-known cases), and gained attention only recently by material science researchers. Some of the most interesting examples of the recent achievement in this field are reviewed in the following.

The most famous example of technology inspired by plants that is commonly used in many different fields and applications is the Velcro invention. Velcro resulted in 1948 from a Swiss engineer, George de Mestral, noticing how the hooks of the plant burrs (*Arctium lappa*) stuck in the fur of his dog. From this observation, George de Mestral derived the idea of a novel type of zip fastener (Velcro, 1955).

Another interesting industrial solution resulting from the study of material properties in plants is Lotusan®, a paint for self-cleaning surfaces. The scientific basis of this invention is the so-called lotus effect: some leaves of water-repellent plants, such as *Nelumbo nucifera* (lotus) and *Colocasia esculenta*, show superhydrophobic and self-cleaning properties, due to the presence of a hydrophobic coating and a multi-scale hierarchical roughness (microstructure formed by papillose epidermal cells covered with epicuticular wax tubules) (Barthlott and Neinhuis, 1997; Bhushan, 2009). The observation that the leaves of the lotus are always clean, despite growing in muddy and stagnant water, led to the production of this new product. Similar surface textures are used to repel dirt or make it easily removed, and have been observed in many other plants and in other systems, such as insect wings (Wagner et al., 1996).

Concerning intrinsic mechanical structural properties, an interesting example of technological solution inspired by plants is given by cellular or sandwich structures consisting of solid shells filled with compliant cellular cores, which have increased resistance against axial buckling while being light weight. These properties are typically required in aerospace applications (Meyers et al., 2008). Particularly interesting are bamboo-inspired fibers and composites for the development of structural engineering materials (Li et al., 1995). Bamboo is one of the strongest natural structural composite materials in which weight to strength ratio is optimized. While having similar chemical composition to wood, bamboo mechanical properties are very different: e.g., tensile strength of bamboo is in the order of 150–520 MPa while in wood it is limited to 30–220 MPa. Such differences are due to the peculiar cellular structure at macro, meso, and micro-scale. At the macro-scale bamboo is a hollow cylinder with many longitudinal nodes, while wood is a solid cylinder. At the meso-scale, bamboo has a non-linear gradient structure in which vascular bundles and thin-walled cells are present, while in wood there are alternated "rings" or layers of spring and summer wood. At the microscale, wood and bamboo fibers have basically analogous multilayer concentric hollow-tubes structure, in which each layer is reinforced with microfibrils. However, the two micro-structures have such differences as their cell wall thickness, layer number, and microfibrillar angle. The relative arrangement of such interlayers and of transition zones in bamboo fibers are known to have a significant effect on fibers properties. By modeling the mechanical behavior of bamboo and by mimicking the arrangement of its fibers, biomimetic fibers realized with composite engineering materials, like glass fiber-reinforced polyester or epoxy resins, have been realized showing improved properties in terms of interlaminar shear strength (Li et al., 1995).

From a structural point of view, another interesting example is given by the pomello, *Citrus maxima*, which shows a peculiar damping system, allowing the fruit to resist to a drop from heights of up to 15 m. In this case the fruit wall, thanks to a complex hierarchical structure consisting of interconnected porous layers combined with branched fiber networks, is able to dissipate more than 90% of the impact energy thus preventing pulp damage (Martone et al., 2010; Thielen et al., 2013). The principles have been exploited in the development of new materials for enhanced crash absorbers, based on metal foams (Fischer et al., 2010).

Bioinspiration from plants is not only restricted to structural materials but also another interesting aspect to study is related to the structure-to-movement relation. For instance, the study of the arrangement of the materials in plant structures and, more in general, of plant biomechanics, as well as of the physical mechanisms used by plants to achieve movements is increasingly recognized as an interesting approach to derive bioinspired devices. The hierarchical structure and orientation of microfibrils of the cellulose layer in the plant cell walls allow passive movements that are mainly driven by changes in environmental conditions. These systems do not require further control and energy supply by the organism once their growth is completed. This plant features make them interesting for biomimetic transfer and offer new insights for designing smart actuators, smart sensors, and new materials, which "respond" to environmental changing conditions (e.g., humidity and temperature).

On these topics, a nice example is given by the Flectofin®, a new façade-shading system that is inspired by the open mechanism of the flower of the Bird-of-Paradise (*Strelitzia reginae*). In this flower, the bending mechanism that expose the nectar to birds that land on the perch is used to transfer the pollen to the near flowers, and it is based on a hingeless movement in which an external mechanical force initiates a complex deformation of multiple structural members. In this case, this mechanism has been implemented by means of a glass fiber-reinforced composite, allowing large elastic deformation, driven by variation of temperature that passively can control the façade-shading systems (Lienhard et al., 2011).

A new actuation mechanism based on reversible adsorption and desorption of environmental humidity was recently proposed (Taccola et al., 2013). This device combines the possibility to achieve active and passive actuation with a single composite material, i.e., poly(3,4-ethy lenedioxythiophene):poly(styrenesulfonate) (PEDOT:PSS), which is a well-known conjugated conducting polymer that exhibits a unique water absorption capability (due to the hydrophilic PSS). Actuation can be obtained by coupling an ultra-thin film of PEDOT:PSS (a thickness of several hundreds of nanometers) with a passive elastomeric layer (a thickness of hundreds of μm) in a bilayered fashion. If the humidity in the ambient air increases, then the PEDOT:PSS layer adsorbs water vapor and increases in volume, which results in bending of the structure due to the constraints of the passive layer.

The osmotic principle, so cleverly harnessed by plants to produce their movements, has also been considered for actuating several artificial systems (Sudaresan and Leo, 2008; Piyasena et al., 2009). In particular, an osmotic actuator was proposed to steer the tip of a mechatronic system inspired by the apex of the plant roots (Mazzolai et al., 2011) for soil exploration and monitoring. Furthermore, a dynamic model of the osmotic actuation concept was developed based on elements that remember key basics in plants, such as an osmotic membrane, an actuation chamber that contains the osmotic agent and both a rigid and a deformable boundary, and a solute reservoir chamber (Sinibaldi et al., 2013).

ROBOTIC SOLUTIONS INSPIRED BY PLANT ROOTS

The distinctive living form that confers a plant root the power to penetrate and explore soil at first instance concentrates in the root apex. Briefly said, root apex grows adding new cells, which move from meristem to elongation zone, where they expand axially because of the water absorbed by osmosis and the directional loosening of the cell wall. This action creates an interface between the soil and the root apex, and allows the root to penetrate the soil with only a small part of its body (the apex), while the remainder of the structure is stationary and in contact with the soil (the mature region). Hairs generating at root fringe anchor the structure to the soil and increase the available surface for water and nutrients uptake. Cell division and morphology are influenced by the surrounding environment, and so the growth process at the tip enables the root to adapt to the environmental conditions such as soil texture and mechanical impedance (Dexter, 1987).

Growth kinematics and morphological features were extensively investigated (Bengough et al., 2011) through observations of living roots in various environmental conditions, together with modeling and simulation (Clark et al., 2008). Cells expansion, elongation, sloughing, and differential elongation are therefore all well-known strategies concentrated at root tip level, that are used to follow low impedance pathways in the soil.

Taking advantage of strategies to penetrate and explore soil as well as to maintain good performance in terms of energy efficiency are totally new research goals in bioinspired robotics that are attracting different scientific communities like biology, physics, and material science. A first multidisciplinary research effort in this direction is focusing on robotic solutions that are called "PLANTOIDS," which are robotic systems equipped with distributed sensing, actuation, and intelligence to perform soil exploration and monitoring tasks.

In such complex process of inventing efficient mechanisms for soil penetration, the first important strategy to be deciphered is related to the way dynamical friction develops on a limited part of the root body (tip level) while the rest is firm and anchored to the soil. An important aspect of growth regards the sloughing of cells from the tip, and to this regard a simple but effective artificial method was investigated (Sadeghi et al., 2013). A soft and flexible tubular skin was stored inside a rigid shaft so that it could traverse the hole to the external surface of the shaft. The outward movement of skin from the tip was activated by a motor on top of the shaft connected to the skin, and it provided a low-friction interface between the shaft and soil. A second aspect was also implemented: to mimic the anchoring achieved by the natural root hairs that grow laterally to its wall, artificial soft hairs were integrated on the external part of the skin in contact with the soil. The robotic system was able to penetrate a granular substrate using the movement of the soft skin without adding any external force. It was demonstrated that upon increasing the hair density, the efficiency of the penetration increased to approximately 30% and the axial penetration force decreased owing to the skin movement.

However, the proposed sloughing system alone is not able to provide friction reduction for an artificial root; externally the skin interaction with soil along the shaft is stationary, but internally the relative movement of the skin and the penetrating shaft can produce a resisting friction force which increases by increasing the depth of penetration. Concentrating more on the root-growth approach at the apical region, the "elongation from the tip" (EFT) strategy was studied with artificial systems (Tonazzini et al., 2013). It is noteworthy to mention that the amount of penetration energy required for a rigid probe with EFT was less than the energy required for the entire artificial root body insertion in granular media (reductions went from approximately 20 to 50%).

On the basis of these results, a system that grows at the apical area by the addition of new material was recently developed (Sadeghi et al., submitted for publication), which represents a totally new approach in robotics, embodying *artificial growth*. The system embeds a *growing zone* and a stationary *mature zone*. The *mature zone* consists of a hollow, tubular structure that allows the transfer of new material from a spool (external to the robotic root) and power to a *growing zone*. The growing zone is based on an additive layering mechanism that generates a force for penetration into the soil and that transfers a material (which is in the form of a filament) from the external system to the head (or tip). The *growing* capability at the tip therefore is achieved by means of an additive manufacturing technique, and the layer-by-layer deposition creates the *mature zone*, which uses a tubular body as a support structure that moves axially inside the mature zone in a passive manner. The new material is distributed on the surface of the deposition head, in the *growing zone*.

From these initial achievements, the basic principles learned from the natural root behavior could be validated and, by adding flexibility, the system could be designed to change the direction and navigate around the obstacles. In Figure 1, a first prototype of the PLANTOID is shown. The system has a main trunk to which two roots are connected. One root embodies artificial growth like explained above and it elongates and penetrates the soil by an additive process of material (Sadeghi et al., submitted for publication). Another root integrates the bending capability in three directions and a sensory system for temperature, humidity, gravity, and touch.

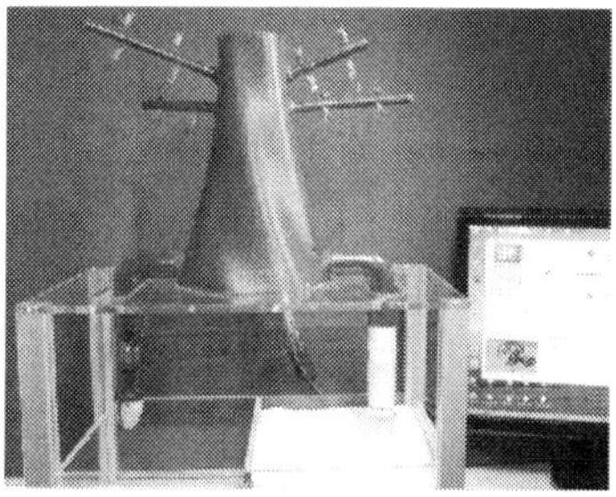

Figure 1. The PLANTOID robot. It integrates two roots that implement the elongation movement and the bending motion with a sensorized tip, respectively. In this prototype, the tip connected to the bending root integrates sensors for touch, humidity, temperature, and gravity. The branches integrate artificial leaves based on PEDOT:PSS material that move in response to changes in environmental humidity.

The approach to take inspiration to Plant Kingdom and to translate plant features in artificial solutions, especially in the robotic field, is moving its first steps. Many of the structural, functional, and physiological properties of plants and plant parts represent a revolutionary source of inspiration because they are based on evolutionary strategies aimed at reducing energy consumption and optimizing the use of local resources. This results, for example, in plant structures that exploit the environmental energy to move or implement efficient motion strategies. Therefore, the first lesson that we must take from plants is to watch them with different eyes.

CONFLICT OF INTEREST STATEMENT

The authors declare that the research was conducted in the absence of any commercial or financial relationships that could be construed as a potential conflict of interest.

ACKNOWLEDGMENTS

The authors acknowledge the financial support of the Future and Emerging Technologies (FET) program within the Seventh Framework Program for Research of the European Commission, under the PLANTOID Project, FET-Open grant number: 293431.

REFERENCES

1. Ayers J., Witting J. (2007). Biomimetic approaches to the control of underwater walking machines. Philos. Trans. R. Soc. Lond. A 365, 273–29510.1098/rsta.2006.1910.
2. Barthlott W., Neinhuis C. (1997). Purity of the sacred lotus, or escape from contamination in biological surfaces. Planta 202, 1–810.4315/0362-028X.JFP-12-021.
3. Bengough A. G., McKenzie B., Hallett P., Valentine T. (2011). Root elongation, water stress, and mechanical impedance: a review of limiting stresses and beneficial root tip traits. J. Exp. Bot. 62, 59.10.1093/jxb/erq350.
4. Bhushan B. (2009). Biomimetics: lessons from nature-an overview. Philos. Trans. A Math. Phys. Eng. Sci. 367, 1445–148610.1098/rsta.2009.0011.
5. Burgert I., Fratzl P. (2009). Actuation systems in plants as prototypes for bioinspired devices. Phil. Trans. R. Soc. Lond A 367, 1541–155710.1098/rsta.2009.0003.
6. Cianchetti M., Arienti A., Follador M., Mazzolai B., Dario P., Laschi C. (2011). Design concept and validation of a robotic arm inspired by the octopus. Mater. Sci. Eng. C 31, 1230–123910.1016/j.msec.2010.12.004.
7. Clark L. J., Ferraris S., Price A. H., Whalley W. R. (2008). A gradual rather than abrupt increase in soil strength gives better

root penetration of strong layers. Plant Soil 307, 235–24210.1007/s11104-008-9602-8.

8. Cutkosky M., Kim S. (2009). Design and fabrication of multi-material structures for bioinspired robots. Phil. Trans. R. Soc. A 367 no. 1894, 1799–181310.1098/rsta.2009.0013.
9. Darwin C. (1880). The Power of Movements in Plants. London: Murray.
10. Dexter A. R. (1987). Mechanics of root growth. Plant Soil 98, 303–31210.1007/BF02378351.
11. Dumais J., Forterre Y. (2012). "Vegetable dynamics": the role of water in plant movements. Ann. Rev. Fluids Mech. 44, 453–47810.1146/annurev-fluid-120710-101200.
12. Fischer F., Thielen M., Loprang R. R., Seidel R., Fleck C., Speck T., et al. (2010). Pummelos as concept generators for biomimetically inspired low weight structures with excellent damping properties. Adv. Eng. Mater. 12, B658–B66310.1002/adem.201080065.
13. Forterre Y. (2013). Slow, fast and furious: understanding the physics of plant movements. J. Exp. Bot. 64, 4745–476010.1093/jxb/ert230.
14. Fratzl P., Barth F. G. (2009). Biomaterial systems for mechanosensing and actuation. Nature 462, 442–44810.1038/nature08603.
15. Gao H., Wang X., Yao H., Gorb S., Arzt E. (2005). Mechanics of hierarchical adhesion structures of geckos. Mech. Mater. 37, 275–28510.1016/j.mechmat.2004.03.008.
16. Hart J. W. (1990). Plant Tropisms and Other Growth Movements. London: Unwin-Hyman.
17. Hawkes E. W., Eason E. V., Asbeck A. T., Cutkosky M. R. (2013). The Gecko's Toe: scaling directional adhesives for climbing applications. IEEE/ASME Trans. Mechatron. 18, 518–52610.1109/TMECH.2012.2209672.
18. Ijspeert A. J., Crespi A., Ryczko D., Cabelguen J. M. (2007). From swimming to walking with a salamander robot driven by a spinal cord model. Science 315, 1416–142010.1126/science.1138353.
19. Kim S., Laschi C., Trimmer B. (2013). Soft robotics: a bioinspired evolution in robotics. Trends Biotechnol. 10.1016/j.tibtech.2013.03.002.

20. Li S. H., Zeng Q. Y., Xiao Y. L., Fu S. Y., Zhou B. L. (1995). Biomimicry of bamboo bast fiber with engineering composite materials. Mater. Sci. Eng. C 3, 125–130.
21. Lienhard J., Schleicher S., Poppinga S., Masselter T., Milwich M., Speck T., et al. (2011). Flectofin: a nature based hinge-less flapping mechanism. Bioinspir. Biomim. 6, 045001.10.1088/1748-3182/6/4/045001.
22. Ma K., Chirarattanon P., Fuller S., Wood R. J. (2013). Controlled flight of a biologically inspired, insect-scale robot. Science 340, 603–60710.1126/science.1231806.
23. Martone P. T., Boller M., Burgert I., Dumais J., Edwards J., Mach K., et al. (2010). Mechanics without muscle: biomechanical inspiration from the plant world. Integr. Comp. Biol. 50, 888–90710.1093/icb/icq122.
24. Mazzolai B., Margheri L., Cianchetti M., Dario P., Laschi C. (2012). Soft-robotic arm inspired by the octopus: II. From artificial requirements to innovative technological solutions. Bioinspir. Biomim. 7, 025005.10.1088/1748-3182/7/2/025005.
25. Mazzolai B., Mondini A., Corradi P., Laschi C., Mattoli V., Sinibaldi E., et al. (2011). A miniaturized mechatronic system inspired by plant roots for soil exploration. Mechatron IEEE/ASME Trans. 16, 201–21210.1109/TMECH.2009.2038997.
26. Meyers M. A., Chen P.-Y., Lin A. Y.-M., Seki Y. (2008). Biological materials: structure and mechanical properties. Prog. Mater. Sci. 53, 1–20610.1016/j.pmatsci.2007.05.002.
27. Pfeifer R., Bongard J. C. (2007). How the Body Shapes the Way We Think: A New View of Intelligence. Cambridge, MA: MIT Press.
28. Pfeifer R., Lungarella M., Iida F. (2012). The challenges ahead for bio-inspired 'soft' robotics. Commun. ACM 55, 76–8710.1145/2366316.2366335.
29. Piyasena M. E., Newby R., Miller T. J., Shapiro B., Smela E. (2009). Electroosmotically driven microfluidic actuators. Sens. Actuators B 141, 263–26910.1016/j.snb.2009.05.014.
30. Sadeghi A., Tonazzini A., Popova L., Mazzolai B. (2013). "Robotic mechanism for soil penetration inspired by plant root," in Proceedings of the 2013 IEEE International Conference on

Robotics and Automation, ICRA2013, Karlsruhe, Germany, 3457–346210.1109/ICRA.2013.6631060.

31. Sinibaldi E., Puleo G. L., Mattioli F., Mattoli V., Di Michele F., Beccai L., et al. (2013). Osmotic actuation modeling for innovative biorobotic solutions inspired by plant kingdom. Bioinspir. Biomim. 8, 025002.10.1088/1748-3182/8/2/025002.
32. Sudaresan V. B., Leo D. J. (2008). Modeling and characterization of a chemomechanical actuator using protein transporter. Sens. Actuators B 131, 384–39310.1016/j.snb.2007.11.057.
33. Taccola S., Zucca A., Greco F., Mazzolai B., Mattoli V. (2013) "Electrically driven dry state actuators based on PEDOT:PSS nanofilms," EuroEAP 2013, International Conference on Electromechanically Active Polymer (EAP) Transducers & Artificial Muscles, Duebendorf.
34. Thielen M., Schmitt C. N. Z., Eckert S., Speck T., Seidel R. (2013). Structure–function relationship of the foam-like pomelo peel (Citrus maxima) – an inspiration for the development of biomimetic damping materials with high energy dissipation. Bioinspir. Biomim. 8, 025001.10.1088/1748-3182/8/2/025001.
35. Tonazzini A., Sadeghi A., Popova L., Mazzolai B. (2013). Plant Root Strategies for Robotic Soil Penetration. London: Living Machines.
36. Tramacere F., Beccai L., Mattioli F., Sinibaldi E., Mazzolai B. (2012) "Artificial adhesion mechanisms inspired by octopus suckers," in Proceedings of the 2012 IEEE International Conference on Robotics and Automation, ICRA2012, Saint Paul, MN, USA
37. Velcro S. A. (1955). *Improvements in or Relating to a Method and a Device for Producing a Velvet Type Fabric* Patent No. 721 338 Switzerland.
38. Wagner T., Neinhuis C., Barthlott W. (1996). Wettability and contaminability of insect wings as a function of their surface sculptures. Acta Zoologica (Stockholm) 76, 213–22510.1111/j.1463-6395.1996.tb01265.x.
39. Zang J., Wang Q., Tu Q., Ryu S., Pugno N., Buehler M., et al. (2013). Multifunctionality and control of the crumpling and unfolding of large-area graphene. Nat. Mater. 12, 321–32510.1038/nmat3542.

CHAPTER 11

WIND ENERGY CONVERSION BY PLANT-INSPIRED DESIGNS

Michael A. McCloskey[1,2], Curtis L. Mosher[1], and Eric R. Henderson[1,2]

[1]Department of GDCB, Iowa State University, Ames, IA, United States of America

[2]Creodyne, LLC, 2410 NE 13th Ct, Ankeny, IA, United States of America

ABSTRACT

In 2008 the U.S. Department of Energy set a target of 20% wind energy by 2030. To date, induction-based turbines form the mainstay of this effort, but turbines are noisy, perceived as unattractive, a potential hazard to bats and birds, and their height hampers deployment in residential settings. Several groups have proposed that artificial plants containing piezoelectric elements may harvest wind energy sufficient to contribute to a carbon-neutral energy economy. Here we measured energy conversion by cottonwood-inspired piezoelectric leaves, and by a "vertical flapping stalk"—the most efficient piezo-leaf previously reported. We emulated cottonwood for its unusually ordered, periodic flutter, properties conducive to piezo excitation. Integrated over 0°–90° (azimuthal) of incident airflow, cottonwood mimics outperformed the vertical flapping stalk, but they produced << daW per conceptualized tree. In contrast, a modest-sized cottonwood tree may dissipate ~ 80 W via leaf motion alone. A major limitation of piezo-transduction is charge generation, which scales with capacitance (area). We thus tested a rudimentary, cattail-inspired leaf with stacked elements wired in parallel. Power increased systematically with capacitance as expected, but extrapolation to acre-sized assemblages predicts << daW. Although our results suggest that present piezoelectric materials will not harvest mid-range power from botanic mimics of convenient size, recent developments in electrostriction and triboelectric systems may offer more fertile ground to further explore this concept.

INTRODUCTION

The concept of wind energy harvesting by artificial plants has been considered in primary literature, patents, grants, and by commercial concerns [1–6]. In these schemes electromechanical coupling is achieved by piezoelectric elements assembled into the faux plant. If modest-sized piezoelectric plants could generate tens of watts (daW) off-grid, charging appliance batteries, they might supplement energy from residential turbines and commercial wind farms, helping to achieve the goal of "20% Wind Energy by 2030" [7]. Here we characterize a novel cottonwood-based piezo-leaf, and we predict size constraints on daW-scale harvesting by synthetic plants.

The rationale underlying fabrication of wind-harvesting plants is multifold. Visual and acoustic impacts of wind turbine farms are widely-shared concerns [1], and artificial plants may be less of a hazard to bats and birds than are wind turbines [8]. In the United States, geographic hotspots of average wind speed and population density are widely separated, and given limitations of the grid, local off-grid harvesting for reduced-scale application suggests a niche for piezo-plants. In previous efforts, Oh et al. (2009) built a miniature wind energy-harvesting tree of ten plastic leaves bearing polyvinylidene fluoride (PVDF) inserts that generated 47 mV peak voltage [3], and Zhang et al. (2014) fashioned a "leaf generator" from PZT nanofibers that produced 820 mV peak voltage in wind of 17 m/s [4]. Li et al (2011) developed a far more effective piezo-leaf with a "vertical flapping stalk" that generated ~ 100 to 300 μW in modest wind [5], and "for practical application" suggested assembly of devices with "hundreds or thousands of the Piezo-Leaves, like ivy, tree and forest." Solar Botanic plans to tap both wind and solar energy from groves of "Energy Trees" sprouting their patented "Nanoleaf" [2]. The VTT Technical Research Center in Finland took this concept a step further by 3d printing of solar energy-harvesting leaves "capable of harvesting kinetic energy from a surrounding environment such as wind" [6].

The patent motion of leaves and branches in wind dissipates wind energy. Given a suitable transduction scheme, mimicry of plant anatomies could yield visually appealing architectures for mid-scale wind energy harvesting. Wind energy that is partitioned among leaves, branches, bole and roots of a tree induces motion at sub-Hz to a few Hz (Results), and in a 10 mph breeze a cottonwood tree with 500,000 leaves may dissipate 80 W via leaf motion alone. If a faux tree scavenged half this power, it could deliver off-grid energy sufficient to charge batteries of small household appliances. Tree leaves in the genus *Populus* flutter in wind speeds at or below the ~ 7 mph threshold of conventional turbines [9], suggesting deployment of faux trees in populous settings where tall, noisy turbines are viewed a nuisance. Cell phone towers camouflaged as trees already have been used for decades in some urban settings, and their functionality would be amplified if piezoelectric leaf assemblages prove capable of harvesting significant wind energy. The topic of large-scale vibration energy harvesting, per se, is treated elsewhere [10].

To adapt piezo transducers to plant structures it is necessary to consider frequencies, stress levels, and stochastic movement *vis a vis* modes that effectively excite piezoelectric generators. Piezoelectric elements act as capacitors that self-charge under stress, and capacitive coupling introduces serious challenges to energy harvesting schemes, including impedance mismatch of source and load, dissipation of low frequency signals by RC filtration, and parasitic capacitance. Unique properties of particular materials, e.g., PVDF, electroactive paper, and PZT also set boundary conditions on empirical models.

Here we analyze power conversion by bioinspired plant designs at different frequencies and load resistance, in controlled and natural wind. Our Kynar-based cottonwood mimics improved upon performance of the "vertical flapping stalk" [5], but scaled to the size of a cottonwood tree, power was << daW. Cattail-inspired models with stacked elements showed progressive rise in power with capacitance, but remained << daW per acre. Although our results discourage focus on piezoelectric schemes, alternative methods like the triboelectric system pioneered by Wang's group [11] may offer a more productive test of wind energy harvesting by botanic mimics.

RESULTS AND DISCUSSION

Motional Frequency and Stress

Excitation frequency is a crucial determinant of power output by piezoelectric elements, due to small energy yields per cycle. Thus it is useful to know the frequency range of plant motions in wind. For quaking aspen (*Populus tremuloides*) the dominant flutter frequency of leaves is ~ 3–5 Hz in wind of 2–4 m/s [9, 12]. Alfalfa stems exhibit an oscillatory decay (~ 1.3 Hz) to neutral position after an initial mechanical displacement [13]. Bending frequencies of alfalfa plants in turbulent wind of 1.1 to 6.9 m/s have been inferred from the power spectra of sun flecks striking under the canopy [14]. A prominent peak occurs at 1–2 Hz in wind speeds of 1.2–6.9 m/s, and at 6.9 m/s the major spectral peak spanned ~ 0.8 to 3 Hz, with some contribution from frequencies up to 10 Hz. Wind stress applied to tree trunks of various species, and the induced trunk excursions, have been measured with strain gauges and high-speed cameras [15]. The power spectral density of trunk sway in *Eucalyptus teretecornus* (red gum) in wind of 17 m/s shows a prominent peak at ~ 0.4 Hz, reflecting a dominant motion at this frequency, and Douglas fir has a strong peak at ~ 0.65 Hz [16]. Upon vigorous agitation by mechanical shakers used at harvest, olive tree frames exhibit normal modes at 20.2 and 37.7 Hz [17]. Short of gale force winds, trunk movements at these frequencies are unlikely to dissipate much wind energy (but see [18]). Thus, wind-driven motion of aboveground plant structures appears concentrated in a band from sub-Hz to a few Hz.

The central role of frequency narrows a search for piezo location to small features like leaves. As well, non-chaotic is a more effective stimulus than is random motion. We modeled cottonwood leaves because their vertically flattened petioles compel blades to flutter side-to-side, concentrating energy into one oscillation mode. A laminated PVDF strip measuring 6.15 x 1.22 cm serving as petiole was attached to a faux leaf blade made of double thickness lamination plastic. In fan-generated wind of steady velocity this construct mimicked the side-to-side oscillations of live cottonwood leaves and had similar threshold wind speed (≤ 3 knots). Fig 1A shows power generated as a function of wind speed, and Fig 1B shows primary data from the rising phase of response. Both frequency and peak voltage increased with wind speed. Flutter frequency measured from slow motion video was ~ half the electrical frequency, as expected with full-wave rectification. Leaf motion became chaotic at ~ 23 knots and nearly ceased at ~28 knots. In the initial, rising phase, power was concentrated in a narrow band that shifted to higher frequencies with wind speed, as indicated in Fig 1C and by the triangles in Fig 1A, reaching ~ 4.5 Hz at 16.1 knots.

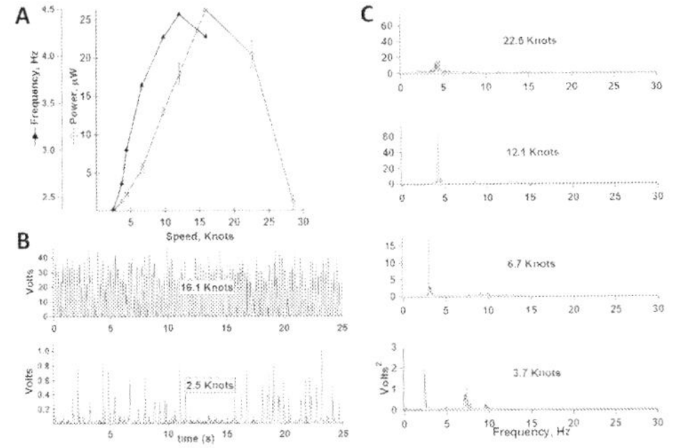

Figure 1. Properties of model cottonwood leaf in fan-generated wind: (A) Power dissipated across 10 MΩ load (circles), and frequency of major band (triangles) as function of wind speed; (B) Raw voltage data at two wind speeds; (C) Power spectra at four wind speeds, showing shift of major band to higher frequencies with increase in wind speed. Leaf was angled 60 degrees downward from horizontal.

Fig 2 shows power produced by the same PVDF strip, sans leaf blade, excited by pulses of nitrogen gas at 34 PSI. Pulsatile air excitation allows more precise control of frequency than does steady airflow. For this piezo, power continued to increase with frequency to at least 20 Hz. It jumped disproportionately at 4 Hz, possibly due to mechanical resonance. Greater power output by the complete leaf model likely derives from a larger proof mass (leaf blade), and a bending moment demonstrably less for the partial system. For all frequencies there was an optimal load resistance, also observed in the experiment below.

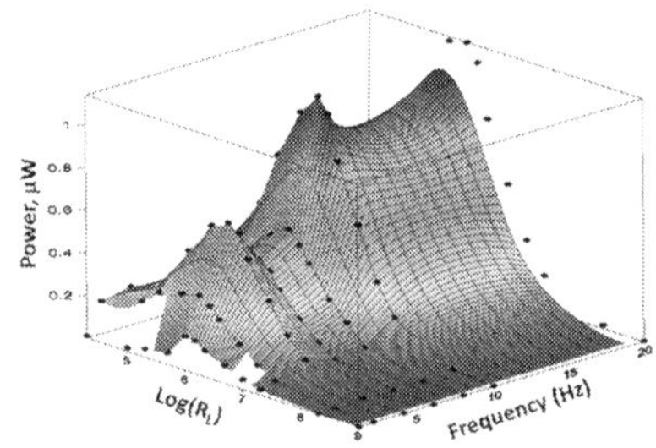

Figure 2. Effect of excitation frequency and load resistance on power output from PVDF strip exposed to repetitive pulses of N_2 gas at 34 PSI. Optimal R_L corresponds to capacitive reactance of piezo. Sweet spot appears to exist at ~ 4 Hz, perhaps due to mechanical resonance.

Fig 3 shows measurements made on a triplex stack of 15.6 x 1.88 cm PVDF strips that was suspended from its base and flexed periodically by pulses of nitrogen gas at ~ 23 PSI. Power rose to a maximum near 4 Hz and decayed sharply above this frequency (not shown). In bending a piezoelectric cantilever, strain (and voltage) is proportional to displacement perpendicular to the film's #1 axis length. At frequencies ≤ few Hz, oak or cottonwood leaf-sized PVDF cantilevers bend and relax a few cm in response to repetitive bursts of gas at ~ 23 PSI. At higher frequencies the response was too sluggish to allow cm-scale displacement and relaxation, though stress from each air burst was greater than the pressure difference (~ 3–30 Pa) across the device due to modest wind (2.2–6.7 m/s).

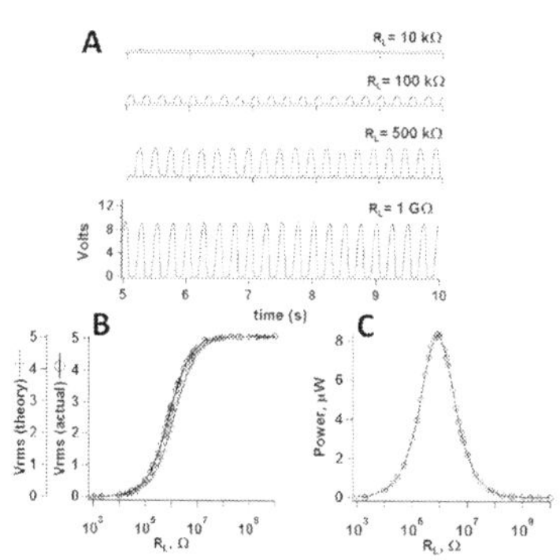

Figure 3. Effect of load resistance on energy conversion by a triplex stack of PVDF elements excited by repetitive pulses of N_2 gas at 4 Hz: (A) Primary data show progressive drop in voltage with R_L. (B) Measured (circles) and theoretical (crosses) RMS voltage drops across load resistance; (C) Effect of R_L on power (V^2/R_L), with optimal harvesting near 1 MΩ.

Impedance Matching

Impedance mismatch of source and load is a huge barrier to wind energy harvesting by piezo-plants. Piezoelectric elements have high output impedance dominated by capacitive reactance, Xc, ($Xc = [2\pi \cdot F \cdot C]^{-1}$), where Xc << ohmic resistance of PVDF [volume resistivity 10^{14}–10^{15}ohm-cm]. At a frequency (F) = 4 Hz, Xc of a single 28 μm thick PVDF element considered earlier is ~ 4 MΩ (C = 10 nF). Fig 3A shows rectified voltage transients in a triple stack of 15.6 cm x 1.88 cm PVDF strips suspended from their base and excited by repetitive bursts of N_2 gas (34 PSI) applied to their tip. Note the decrease when R_L was reduced from 1 GΩ to 10 kΩ. The load resistance in series with impedance of the piezo forms a voltage divider, and the fraction of source voltage realized across load is $R_L/(R_L + Xc)$. Maximal voltage is observed at $R_L >> Xc$. Fig 3B shows reasonable agreement between measured and theoretical RMS voltage over a broad range of R_L. Voltage was maximized beyond ~ 100 MΩ. In principle, the optimal efficiency of power transfer to load occurs for $R_L = Xc$. Fig 3C shows that with a PVDF piezo approximately the size of a cottonwood leaf, optimal efficiency of energy transfer to load occurred in the MΩ range, near Xc. A rechargeable Li^+ ion or NiMH battery has internal resistance < 1 Ω, so for direct charging, transfer efficiency < 2.5 x 10^{-5}%.

Harvesting efficiency can be increased by impedance matching through several methods, including buck boost converters and synchronized switch harvesting on inductor (SSHI) or capacitor (SSHC) [19]. Synchronized switch harvesting is effective for periodic signals like those induced by engine vibration, and the stereotyped flutter of *Populus* leaves might be amenable to SSH. But at 20% efficiency—a liberal estimate, maximum power is several orders too small to support production of daW by devices the size of a large tree.

Charge Production

Charge production is a major intrinsic constraint on mechanical-to-electrical conversion by piezo devices. Peak currents can be amplified by increasing the electrode-coated area of a piezo, or by parallel wiring of a stack of mechanically synced elements. However, the benefit of increased area can be offset by parasitic capacitance, likewise, stacking reduces compliance. To test if a useful compromise can be met in faux plants we modeled cattails (e.g., *Typha latifolia*) which move somewhat like cantilever beams, more uniformly along their length than does a deciduous oak leaf. Fig 4 shows power produced by a stack of 10 PVDF elements attached to the base of a simplified model comprised of a plastic ruler 19" long and flexed at 1.3 Hz by a gated airstream at 11 knots. Capacitance was increased step-wise by parallel connection of successively greater numbers of elements out of ten in the stack. At R_L = 1 MΩ, power increased with capacitance, especially obvious at C ≥ 70 nF, where motional frequency (1.3 Hz) was close to the corner frequency of the RC filter. At C = 100 nF, the model leaf produced 9 μW at 1.3 Hz, a higher frequency than living cattails normally experience (direct observation). At ~ 500 *Typha latifolia*plants per acre and an average of 5.5 leaves per plant, power conversion scaled to 0.025 W/acre.

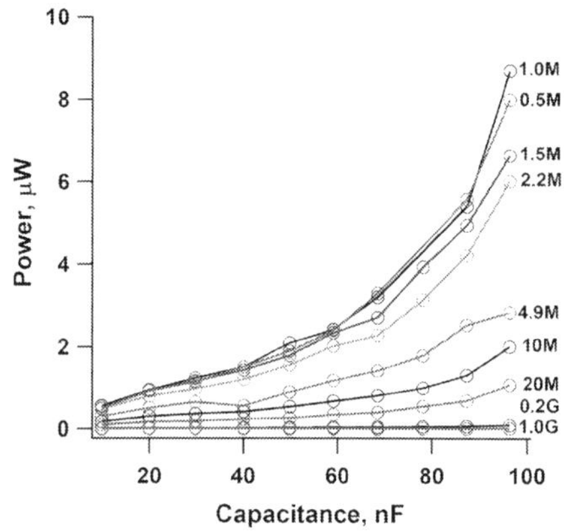

Figure 4. Power generation by a stack of 10 PVDF elements attached to base of 48 cm cattail-inspired model, flexed at 1.3 Hz by a gated air stream (11 knots). Individual elements connected in parallel to increase C step-wise. Approximate ten-fold increase in power with ten-fold increase in C. For R_L = 1 MΩ, Fc = 16.7, 2.3 and 1.7 Hz at 1, 7 and 10 elements connected in parallel.

In separate experiments, freshly-harvested cattail leaves were fitted with PVDF strips, and currents recorded across a 10 Ω shunt resistor (Methods). Leaves were bent by airflow from a reciprocating fan (~1 Hz) pushing air at 10 knots, and by manual displacements (~ 2 Hz). Duplex and triplex stacks of 15.6 x 1.9 cm elements were attached directly within the leaf axil or 21.5 cm above it. Interestingly, the duplex attached 21.5 cm above the leaf axil generated RMS current similar to the double stack located in the leaf axil, with components at higher frequencies than for the axil.

To consider the potential performance enhancement of increasing transducer size, envision stacks of 10 PVDF elements described above (15.6 cm x 1.88 cm; C = 10 nF each) but now expanded 25-fold lengthwise and 4-fold in width. The polarization axes are aligned, elements wired in parallel and the stacks are aligned with the long axis of a tree trunk. If each cycle of trunk sway stresses stacks similar to a 10 knot wind driving the 15.6 cm x 1.88 cm element at 4 Hz, a naïve assumption, then at a trunk frequency of 0.5 Hz each stack could generate 1 mW. To generate 80 W would require 80,000 transducers, ignoring inevitable power loss from impedance mismatch or parasitic capacitance. The increased piezo area would lower capacitive reactance to 32 kΩ, still 10^3 fold greater than internal resistances of batteries (below). The corner frequency of the high pass RC filter in series with a 1 Ω load resistance (5–10 batteries) is ~ 16 kHz, a few thousand times faster than plants move in wind. It might require a Giant Sequoia to accommodate 8 x 10^4 such transducers, certainly not a convenient backyard plant.

Tree Size Projected From Capacitor Energetics

Piezoelectric transducers behave as capacitors that self-charge under stress. Energy stored across a capacitor is $E = ½ CV^2$, with C the capacitance and V the voltage. A 28 μm-thick PVDF sensor 15.6 x 1.88 cm with Ag electrodes (and unilateral 127 μm Mylar backing) has C ~ 10 nF. When subjected to stress similar to that of 10-knot breeze it readily produces open circuit voltage ~ 20 V. At 20 V, energy stored per cycle is ~ 2 μJ, and at a frequency of 4 Hz (natural flutter), max power is ≤ 8 μW. For simplicity this ignores frequency-dependent filtering, and overestimates power. How many such capacitors (leaves) would it take to generate 80 W to 1 kW in a 10 knot wind? Assuming 100% transfer to load and all leaves constantly fluttering—naïve assumptions, 10 M to 125 M synthetic leaves would be required. Apart from cost, the environmental impact would be non-negligible and the spatial footprint in excess of a residential turbine generating equal power. At 10 knots wind power density is ~ 6 W/sq. ft. If leaves sweep an area 5 times greater than their own, efficiency is ~10^{-4}%.

Material Constraints

As noted above, the utility of small piezo devices for wind energy harvesting is handicapped by small charge per cycle. The piezoelectric charge coefficient d_{3n} for PZT is 5-fold larger than that of PVDF (1.1 x 10^{-10} vs. 0.23 x 10^{-10} Coulomb/Newton) [20]. This potential advantage is offset by a 21-fold lower voltage coefficient (g_{31}), 4.2-fold higher density, and 10–20 fold lower compliance than that of PVDF. It would be futile to construct synthetic leaves 4 times heavier and 10 to 20-fold stiffer than real leaves. Macrofiber composites (MFC) of PZT are less dense and more compliant than is conventional PZT, and in our cattail-inspired model a single MFC (C = 185 nF) produced as much power as did the PVDF stack (98 nF) under the same conditions. However, stacking of MFC is infeasible beyond ~ 2 or 3 elements, as MFC are less compliant than biological leaves or petioles (see, e.g., [21], [22]). The elastic modulus of oak leaves and water lily petioles is 20–50 MPa, whereas Young's moduli of PZT MFC are > 15 GPa [23], 10-fold larger than elastic moduli of sunflower petioles [24].

No flexible materials of which we are aware, including cellulose-based EAPap (max d_{3n} 0.07 x 10^{-10} C/N) [25], has d_{3n} values orders greater than that of PVDF or PZT. In one crystal orientation the perovskite $Pb(Zn_{1/3}Nb_{2/3})O_3$–$PbTiO_3$ has a maximal piezoelectric charge coefficient (d_{33}) of 2500 pC/N [26], ~ 20- and ~ 100-fold greater, respectively, than those of PZT and PVDF. A 100-fold boost is inadequate to validate piezo-plant schemes. The protein prestin, a piezoelectric actuator of sound amplification in the cochlea, has a maximum sensitivity of 20 μC/N, 10^4 greater than that of the perovskites [27]. Were a prestin construct expressed in bionic plants, in accessible anatomic sites, it would be a formidable challenge to couple membrane potential changes to charging of batteries.

Electromechanical Coupling Factor

The conversion of mechanical to electrical energy by a piezo has a maximum efficiency characterized by k^2, where k is an electromechanical coupling factor. The magnitude of k varies with excitation mode and material, but for PVDF excited in bending mode, k_{31} = 12%, i.e., PVDF has a theoretical limit of 1.4% for converting mechanical to electrical energy. In theory, induction-based turbines are capable of ~ 70%, and in practice they average 20–30%. PZT and $BaTiO_3$ ceramics have k_{31} values ~ 30% and ~ 21%, respectively, corresponding to 9% and 4% conversion efficiencies [20]. This theoretical advantage over PVDF is counterbalanced by other material properties noted above.

Summation over Leaf Population, Outdoors

On a living tree in wind, a sizeable portion of leaves can sit in a wind shadow created by dense foliage on the upwind side. Also, in a stiff breeze many fully-exposed leaves can remain pushed to an extreme, immobile position for long periods. We asked how power sums over single leaves in our piezoelectric cottonwood model. A primitive piezoelectric tree was fashioned from ten faux cottonwood leaves mounted on a trellis (Fig 5), and power was measured outdoors in natural wind. Whole-tree power at 10 knots was ~ 10-fold greater than for one synthetic cottonwood leaf indoors at 10 knots (Fig 6 vs. Fig 1). Here, average wind speed was estimated as the mean of highest and lowest speed during the 25 s recording. For this small trellis with well-spaced leaves, its plane perpendicular to wind direction, summation was as predicted for parallel wiring of rectified outputs. As the complexity of the faux tree goes up, and leaves are attached at a density and symmetry more typical of living trees (3d vs 2d), an optimal power is realized with less than 100% of leaves contributing.

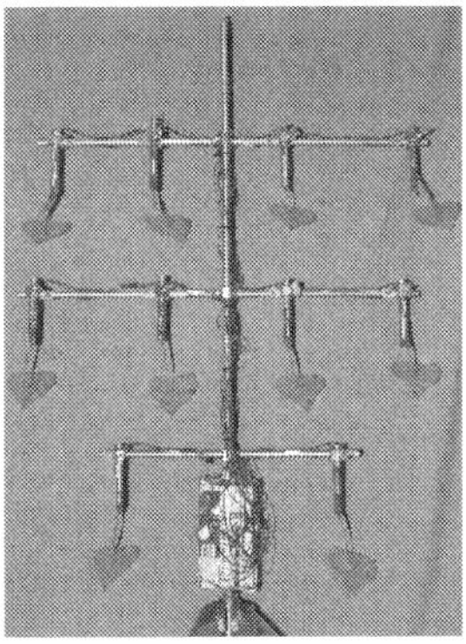

Figure 5. Cottonwood-shaped plastic leaves mounted on aluminum trellis. Output from each Kynar-based petiole was rectified before summation over leaf population.

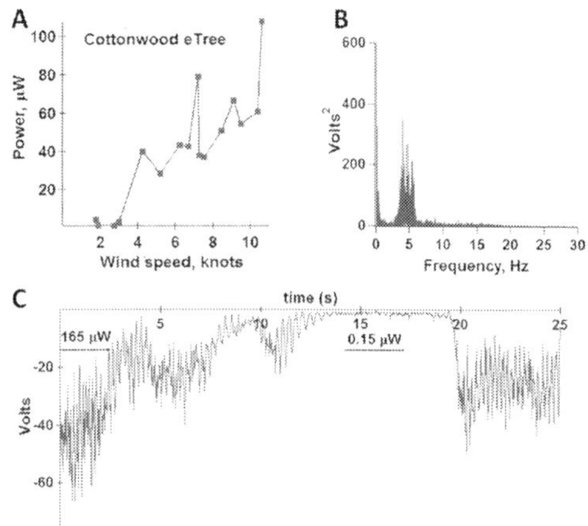

Figure 6. Power output by faux cottonwood trellis as function of wind speed. (A) Average wind speed estimated from mean of highest and lowest speed during the 25 s recording period. At ten knots, power was ~ ten-fold that of single cottonwood mimic indoors. (B) Power spectrum of single leaf during 500 s, in wind fluctuating from 7.5 to 9 knots. (C) Primary voltage data show variation due to fluctuating wind speed and direction.

Airflow in Wind Tunnels vs. Outdoors

Previous contributions tested piezoelectric leaves indoors in wind tunnels (3–5, 28). Optimization of a system using laminar wind of constant velocity does not yield a system optimized to function in natural wind. In nature, wind gusts, changes direction, and is rendered turbulent by neighboring leaves and branches within the plant. Fig 6B is a power spectrum of a single cottonwood leaf on the faux tree, in actual wind. It showed a similar flutter frequency to natural cottonwood leaves despite its larger-than-life petiole. The outsized petiole may help explain why flutter was more sensitive to changes in wind direction than are leaves of authentic cottonwood trees. Our cottonwood model outperformed the vertical flapping stalk (Fig 7) in this regard, but was not as versatile as living cottonwoods; i.e., viable leaves manifest flutter over a broader range of angles of incidence. The nature of leaf attachment to, and mobility of, supporting branches is a crucial determinant of leaf motion in natural wind.

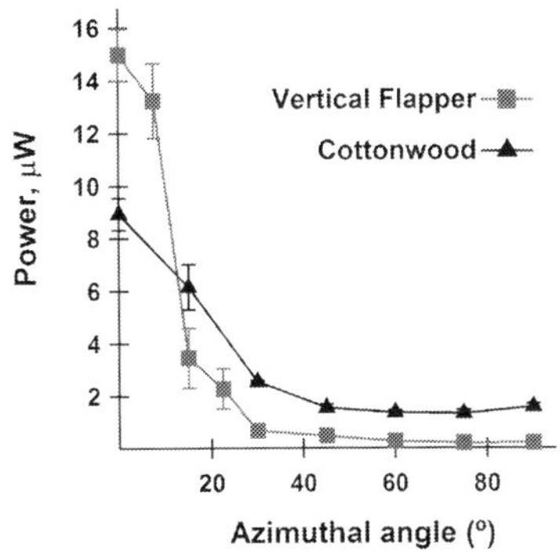

Figure 7. Directional sensitivity of power output from vertical flapping stalk and cottonwood leaf models. Power was maximal for both systems when PVDF petiole was oriented edge-on into wind (8.9 vs. 7.5 knots for flapper vs. cottonwood) and decayed with increase in azimuthal angle. Max power was greater but decayed more steeply for vertical flapping stalk. R_L = 10 MΩ both systems.

Selected Previous Work

Priya's group constructed tiny windmills coupled to piezoelectric bimorphs that produced 5 mW across a 20 kΩ load in a 10 mph airstream [28]. Windmills convert airflow to periodic motion effective for piezo-based harvesting. To harvest ≥ daW by mimicking often chaotic and low frequency motions of plants is a more complex challenge. A decade ago Rick Dickson inspired discussion of wind energy harvesting by piezo-plants (Dickson, R.M., personal communication). Unfortunately, his approach of "…utilizing large arrays with piezoelectric cells connected in series…" was inherently flawed. Series or parallel multiplexing of piezo elements is possible when elements are excited in-phase, but connecting large arrays of independent elements in series yields zero net voltage.

The British company Solar Botanic patented a "Nanoleaf" that captures sunlight and ostensibly, wind energy [2]. A scaling issue suggests the futility of wind energy harvesting by Nanoleaves: Their "millions and millions of pW," nowhere actually measured and published, equates with μW per leaf. Allowing an average of 10 μW per leaf and 500,000 leaves—a substantial tree, power from wind energy would be ≤ 5 W.

In an ingenious construct Li et al (2011) fashioned a piezo-leaf that produced "high" power at low frequency [5]. Their vertical flapping stalk is a PVDF petiole hanging vertically and attached to a triangular plastic blade, the planes of piezo and leaf blade parallel to each other and perpendicular to ground. Under ideal conditions, i.e., with the PVDF strip oriented perfectly edge-on into wind, downstream of a bluff body, it produced more power (100–200 μW across 10 MΩ) than our cottonwood mimic. We modeled their system from available information, and sans bluff body it produced 15–100 μW at wind speed of 8–18 knots (Fig 7). Video records showed high amplitude flexion and torsion, which likely explain high power but also distressing noise (65–80 dB at 10 knots).

We asked how output under ideal conditions compared to that at different azimuthal and altitudinal angles, as present in wind gusts outdoors. The faux cottonwood showed striking dependence on altitudinal angle. In 8 knot wind it generated 0.2, 22, 8.5 and 9.9 μW at angles of 0, -30, -60 and -90° from horizontal. Although not addressed by Li et al., we found that the flapping stalk stopped flapping for any but a small range of azimuthal angles nearly parallel to wind direction (Fig 7). Integration over 90 degrees gave 273 vs. 221 μW-degree for cottonwood vs. vertical flapper. Although the flapper beat the cottonwood in max power, the cottonwood capitalized on changes in wind direction better than did the flapping stalk.

A bluff body boosted power from the flapping stalk 10-fold under idealized wind tunnel conditions [5]. Bluff bodies also were used in wind energy harvesting by a simpler system yielding sub-μW power [29]. Clearly, a static bluff body placed upstream is of no value when wind shifts direction, blowing from leaf towards bluff body. Fig 6C shows output from a trellised cottonwood, outdoors in natural wind. During 25 s, as wind changed direction and speed, power ranged from 0.15 to 165 μW. We conclude that modest changes in wind direction, and turbulence caused by nearby leaves and branches, compromise performance of e-Trees in ways not considered in previous work. Further, to employ the vertical flapping stalk or other piezo-leaves in real-world settings would require large collections of leaves. Yet to assemble >10,000 flapping stalks or cottonwood mimics, each leaf occasionally producing 100 μW on a windy day, would yield ≤ 1 W. By comparison, a solar panel having much smaller footprint can average 200 W on a sunny day. An alternate mechanism of energy transduction, or electroactive materials of orders greater energy density than present ferroelectrics, may place the piezo-plant idea on firmer ground.

Conclusions

Our findings indicate that generation of ≥ daW from piezoelectric botanic mimics of practicable size is not a near-term reality. Transduction schemes of greater energy density such as triboelectric conversion [11] or perhaps electrostriction of flexible polymers [30] may be worthy of testing in synthetic plants. Although valid concerns attend displacement of living with artificial plants, to demonstrate mid-range power for off-grid use in a modest-sized plant mimic would provide strong impetus for real world testing and possible development of such devices.

MATERIALS AND METHODS

Several types of PVDF sensors were donated by or purchased from Measurement Specialties, Inc. (Hampton, VA). Macrofiber composites of PZT (d31-MFC) were purchased from Smart Material Corporation (Sarasota, FL). Resistors, capacitors, diode bridge rectifiers, op amps and other electronic components were from Newark (Gaffney, SC) or Digi-Key (Thief River Falls, MN). Capacitance was measured with a BK Precision 890B capacitance sorter. Airspeed was measured with a hand-held anemometer (BK Precision #731A).

In some experiments (Fig 3) voltage measurements were made with a USB digital scope and stimulator (Stingray DS1M12) from EasySYNC Ltd. (Hillsboro, OR). Digitized data from the Stingray interface were exported from EasySYNC into Excel. In most experiments a Tektronix digital storage scope (TDS 2002C) was used with Open Choice Desktop, and data exported to Excel. IGOR Pro (Wavemetrics, Lake Oswego, OR) was used to analyze and plot all data.

Piezo output was rectified with diode bridge rectifiers (Multicomp AM154 and Multicomp W01), the signal passed through variable load resistance, and RMS voltage across R_L measured over periods of 25 s to 9 m. Power was estimated as $[Vrms]^2/R_L$. In living cattail experiments, currents were measured from the voltage drop across a 10 ohm precision resistor, using an LTC 1050 op amp with feedback resistance such that gain = 10^4. A capacitor of similar capacitance to the piezo was used in a few trials to smooth the signal. For a cattail mimic, ten DT4 elements (MSI part # 1-10022149-0) were attached back-to-back with their electrical axes aligned and electrodes wired in parallel with 3M™ EMI Shielding

Tape #1183. Each piezo was addressable individually. 3M VHB adhesive transfer tape (F9460 PC) was used to bond elements to each other and to a 12 inch plastic ruler (Wescott B-70) extended to 19" with beavertail sail. DT4 elements are ~ 40 μm thick (PVDF = 28 μm) and VHB tape ~ 50.8 μm. Cattails are not perfect cantilevers, as they often are twisted ≥ one revolution lengthwise, and their cross sectional shape and area vary lengthwise, the distal ~ ¼ of leaf more flexible and strained by wind than the proximal 1/3. However, to capitalize on this tip geometry is difficult because piezo stacks here limit strain. The number density of cattail plants in a local stand was measured by counting individuals in two 100 sq. ft. domains.

A rotary fan at constant speed and wind direction and for higher speeds a heavy duty dust collection system (Penn State Industries, model DC-3XL) were used to induce flutter of cottonwood mimics. In some experiments a pneumatic PicoPump (World Precision Instruments, Sarasota, FL) was used to deliver repetitive bursts of nitrogen gas at 23 or 34 PSI to the tips of elements anchored at their base. Gating frequency was controlled by TTL output from the USB scope. Artificial cottonwood leaf blades were fashioned from double-layered lamination plastic or 160 μm Tyvek™ on a laser cutter programmed with authentic cottonwood leaf template. Of several materials tested, Tyvek and lamination plastic allowed the most consistent side-to-side motion of the PVDF petiole at low wind speeds, mimicking the flutter of natural cottonwood leaves. The petiole was constructed from a laminated PVDF element (MSI #LDT2-028K/L). An earlier version with larger blades (13 x 14 cm), and a 15.6 x 1.88 cm piezo-petiole produced much less power than the smaller device.

A synthetic cottonwood tree was made from an aluminum trellis and plastic leaves anchored to the branches with duct tape (Fig 5). Leaves were aligned at 0° (azimuthal) to wind direction, and ~ 45° down from horizontal. Rectified output from each leaf was connected to a parallel bus and voltage measured across 10 MΩ load.

ACKNOWLEDGMENTS

This work was supported by an NSF-SBIR contract (# 1013963) to McCloskey and Henderson. Several types of PVDF elements were gifted by Don Halvorsen of Measurement Specialties, Inc. (Hampton, VA).

AUTHOR CONTRIBUTIONS

1. **Conceptualization:** EH MM CM.
2. **Formal analysis:** MM CM.
3. **Funding acquisition:** EH MM.
4. **Investigation:** MM CM.
5. **Methodology:** MM CM EH.
6. **Project administration:** MM EH.
7. **Resources:** MM CM EH.
8. **Software:** CM.
9. **Supervision:** MM EH.
10. **Validation:** MM CM.
11. **Visualization:** MM CM EH.
12. **Writing – original draft:** MM.
13. **Writing – review & editing:** MM CM EH.

REFERENCES

1. Hadhazy A. Power Plants: Artificial Trees That Harvest Sun and Wind to Generate Electricity. Scientific American. 2009.
2. Van der Beek A. Nanoleaf. Google Patents; 2011.
3. Oh SJ, Han H. J. , Han S. B. , Lee J. Y. , and Chun W. G. Development of a tree-shaped wind power system using piezoelectric materials. International Journal of Energy Research. 2010;34:431–7.
4. Zhang G, Chen, V. , Xu, W. , Galos, R. , Zhou, L. and Shi, Y. , Piezoelectric leaf generator for wind energy harvest. ASME Proceedings; 2014.
5. Li S, Yuan J, Lipson H. Ambient wind energy harvesting using cross-flow fluttering. Journal of Applied Physics. 2011;109(2):026104.
6. VTT Technical Research Center (Finland). 3D printed solar energy trees 2015 [updated 28 Feb 2015]. Available from: http://www.alternative-energy-news. info/3d-printed-solar-energy-trees/.
7. United States Department of Energy. 20% Wind Energy by 2030. 2008.
8. Curry A. Ecology. Deadly flights. Science. 2009;325(5939):386–7. pmid:19628835
9. Roden JS, Pearcy RW. Effect of Leaf Flutter on the Light Environment

of Poplars. Oecologia. 1993;93(2):201–7.

10. Zuo L, and Tang X. Journal of Intelligent Material Systems and Structures. 2013;24(11):1405–30.
11. Wang S, Lin L, Wang ZL. Nanoscale Triboelectric-Effect-Enabled Energy Conversion for Sustainably Powering Portable Electronics. Nano Letters. 2012;12(12):6339–46. pmid:23130843
12. Roden JS, Pearcy RW. Photosynthetic Gas-Exchange Response of Poplars to Steady-State and Dynamic Light Environments. Oecologia. 1993;93(2):208–14.
13. Doare O, Moulia, B. , and de Langre, E. Effect of plant interaction on wind-induced crop motion. Transactions of the ASME. 2004;126:146–51.
14. Tong H, and Hipps L. E. The effect of turbulence on the light environment of alfalfa. Agricultural and Forest Meteorology. 1996;80:249–61.
15. James KR, Kane B. Precision digital instruments to measure dynamic wind loads on trees during storms. Agricultural and Forest Meteorology. 2008;148(6–7):1055–61.
16. Moore JR, Maguire DA. Simulating the dynamic behavior of Douglas-fir trees under applied loads by the finite element method. Tree Physiology. 2008;28(1):75–83. pmid:17938116
17. Castro-Garcia S, Blanco-Roldan G. L. , Gil-Ribes J. A. , and Aguera-Vega J. Dynamic analysis of olive trees in intensive orchards under forced vibration. Trees. 2008;22:795–802.
18. Harne RL, Sun A, Wang KW. Leveraging nonlinear saturation-based phenomena in an L-shaped vibration energy harvesting system. Journal of Sound and Vibration. 2016;363:517–31.
19. Nechibvute A, Chawanda A. , Luhanga P. , and Akande A. R. Piezoelectric energy harvesting using synchronized switching techniques. International Journal of Engineering and Technology. 2012;2(6):936–46.
20. Measurement Specialties, Inc. Piezo Film Sensors Technical Manual. Norristown, PA: Measurement Specialties, Inc. ; 1999.
21. Saito T, and Terashima I. Reversible decreases in the bulk elastic modulus of mature leaves of deciduous Quercus species subjected to

two drought treatments. Plant, Cell and Environment. 2004;27:863–75.

22. Treitel O. Elasticity, Plasticity, Water Content, and Hardness of Plant Tissues of Different Ages. Transactions of the Kansas Academy of Science. 1949;52(2):219–31.
23. Williams RB, Schultz M. R., Hyer M. W., Inman D. J., and Wilkie W. K. Nonlinear Tensile and Shear Behavior of Macro Fiber Composite Actuators Journal of Composite Materials. 2004;38(10):855–69.
24. Hernández LF. Leaf inclination and light interception in the sunflower (Helianthus annuus L.). Importance of the petiole's mechanical and structural properties. 6th Plant Biomechanics Conference. 2009.
25. Kim J, Yun S, Lee SK. Cellulose smart material: Possibility and challenges. Journal of Intelligent Material Systems and Structures. 2008;19(3):417–22.
26. Park S-E, Shrout TR. Ultrahigh strain and piezoelectric behavior in relaxor based ferroelectric single crystals. Journal of Applied Physics. 1997;82(4):1804–11.
27. Dong X-x, Ospeck M. , and Iwasa K. H. Piezoelectric reciprocal relationship of the membrane motor in the cochlear outer hair cell. Biophysical Journal. 2002;82(3):1254–9. pmid:11867442
28. Myers R, Vickers M, Kim HS, Priya S. Small scale windmill. Applied Physics Letters. 2007;90(5):054106.
29. Ovejas VJ, Cuadras A. Multimodal piezoelectric wind energy harvesters. Smart Materials and Structures. 2011;20(8):085030.
30. Wang L, Lallart M, Petit L, Lebrun L, Cottinet P-J, Guyomar D. Low-cost charge of electrostrictive polymers for efficient energy harvesting. Journal of Intelligent Material Systems and Structures. 2015;26(16):2123–36.

CHAPTER 12

BIONICS: BUILDINGS, BRIDGES AND TOOLS

Shiv Sanjeevi Sripathi[1] and Prerna Pandey[2]

[1]Shiv Sanjeevi Sripathi completed his BSc in Biotechnology from Mumbai University and secured second rank in Mumbai University and first rank in his college: Kishinchand chellaram College. Shiv secured first rank in his college KET's V.G.Vaze College and has worked on a stem cell project at the Specialized Centre for Cell Based Therapy (SCCT), KEM Hospital at Mumbai, following which he was a Junior Research Fellow at Microbiology & Cell Biology Department, Indian Institute of Science, Bangalore from August 2009 to November 2011. The field of work involved cloning of cell wall genes and transcription factors in E. coli & M. smegmatis. This was followed by a stint as a Research and Development Executive at Transasia Biomedicals, Mumbai where he worked on patent writing and developing immunoassays from 2011 till 2013. In 2013, he entered the teaching field where he has been tutoring biology and chemistry to the students of 11th and 12th grades along with English grammar and communication skills to people of all age groups.

[2]Dr Prerna Pandey, a biotechnologist with several years of wet lab research experience, worked at International Center for Genetic Engineering and Biotechnology, New Delhi. Her field of research in PhD included isolation and molecular characterization of geminiviruses, genome sequencing, gene annotation, and gene silencing using the RNA interference technology. She has also worked at Transasia Biomedicals and Advance Enzyme Technologies as a scientist. Prerna has published papers in peer-reviewed journals, and has submitted a number of annotated Geminiviral genome sequences in the GenBank, including two novel ones. She has also completed her editing and proof-

reading courses from Society for the promotion of Editors and Proof-readers, UK and now works as a freelance scientific editor and writer. When Prerna is not busy with her assignments, she loves to explore historical places.

INTRODUCTION

During the process of evolution, several revelations and discoveries were seen by man that led to the development of many inventions, most of which have stood the test of time. The usage of the information available in nature is a key to achieve smart solutions. These answers could be applied to several applications and in the design of technology in the form of bionics. Bionics explores technical transformations and applications of underlying principles, procedures, developments, and phenomena working in nature (Lothar Walter et al., 2011).

Bionic building materials refer to substances that imitate the structure, chemical composition, color and ecosystem of living creatures' bodies and are used to design and manufacture new materials that satisfy the human need for form and function (Jiang Dongqing, 2003).

BIONICS AND BUILDINGS

The first instances in using bionics in architecture were seen in the nineteenth century. Paxton based on the studies on huge lotus leaves examined circular shelving and radial structure. This resulted in a model for light glass ceilings (Mehdi Sadri et al., 2014).

The theme of bionics in buildings looks at the remodeling of individual houses, clusters of houses, urban structures etc.

Usually constructions are made to last the test of time, however, the models get redundant with the ever evolving and changing environment. An additional aspect is the usage of resources by the construction sector: for instance, one third of the energy and material consumption and one third of the emissions is credited to the construction business in Germany.

The amount of pollution generated by the industry is also significant with 40% waste in Germany generated by the sector.

These factors all together warrant the application of new ideologies in the arena of construction. The use of the principles of nature in the arena of housing is being pursued as a novel field. Bionics offers several advantages such as:

- Flexibility in the use of buildings: This can be seen with changing the structures of the buildings in accordance with the need and the people of the house.
- They also minimize the use of resources.
- Another aspect is a boost in the living quality and sustained living (Susanne Giesecke and Sabine Hafner-Zimmermann. 2003-2004).

More buildings are expected to be constructed with a key feature being flexibility. This is of utmost importance as the global climate is constantly evolving. Apart from the maintenance, the kinetic challenges such as bending can search for its solutions in bionics (Andreas Hammer, 2016).

According to the principles of bionics, existing bionic architecture is classified into four types:

- Arch form structure: These structures use the design of spinal column of animals. Apart from its solid nature, such designs are reported to be economical.
- Thin-shell structure: members of crustacean families and skulls are used. The design has a mobile approach with an element of even distribution of forces.
- Puffing structure: uses plant and animal cells especially for tents, restaurants in open country, gymnasiums.
- Spiral structure: uses leaf of the plantain that uses principle of sunlight regulation, giving the structure copious amount of sunlight.

Several organisms offer inspiration to the construction of buildings that include: skeletons that offer hardness, wood pecker's skull that is reported as resistant to earthquakes, shells that can bear tremendous amount of pressure etc.

Two categories of architecture and bionics have been described that include:

- The architecture that imitate the anatomy and functions of organisms

- The architecture that imitate a structure's appearance and functions (CHEN FEI and SHA SHA, 2005).

CASE STUDIES IN THE USE OF BIONICS IN BUILDINGS

The Waterloo train station was designed by Nicolas Grimshaw took its inspiration from the structure of fingers in a human hand.

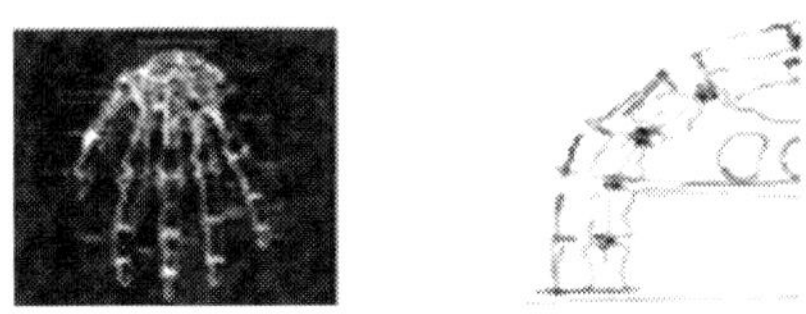

Human digits and Waterloo station: (Images courtesy: Mehdi Sadri et al., 2014)

The reports of dome shaped constructions being inspired from nature can also be seen. Based on structures such as egg shells, their shape along with several layers in the shell with their flexibility offers several interesting advantages.

The construction of skyscrapers has been attributed to inspiration from nature. The stalks of several trees are reported to resist winds and other elements of nature (Billington David, 1983 and Podborschi V. and Vaculenco M, 2002).

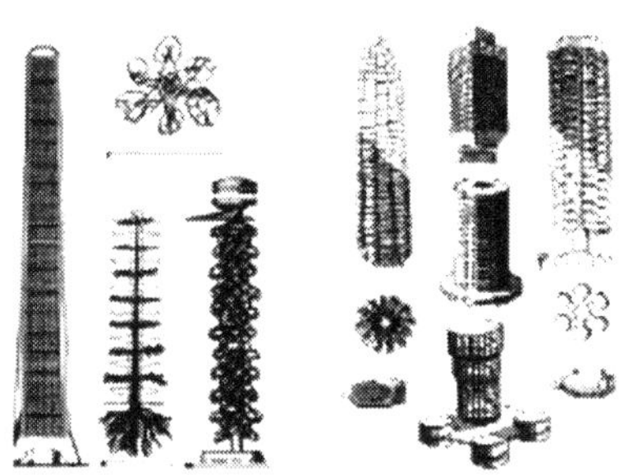

Bionics and skyscrapers: (Images courtesy: Mehdi Sadri et al., 2014)

Spider webs offer several advantages such as their strength as well as elasticity. In order to withstand pressure, the spider and its prey serve as pressure while the web is a fiber mesh. Similarly, the stadium in Munich is constructed on similar lines with elastic strings made of steel and a steel pin that withstands pressure.

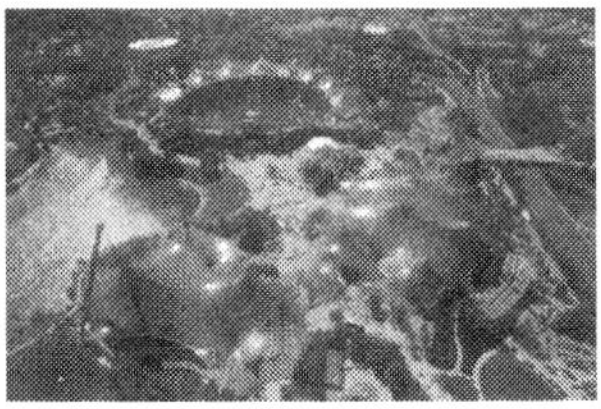

Munich Olympic stadium modeled on a spider web: (Images courtesy: Mehdi Sadri et al., 2014)

The central Swiss Re in London has been modelled using the idea of a sea sponge that possesses a geometric skeleton. This bionic inspired building offers advantages as resistance to wind and aesthetic beauty.

Central Swiss Re in London modeled on a sponge: (Images courtesy: Mehdi Sadri et al., 2014)

The design of a suspended bridge by Isambard Brunel for the Royal Albert bridge near plymouth in England used bionics in using the fossil shape of a dinosaur: Bronotosaurus (Aldersey Williams, 2003 and Salvadori, M, 1990).

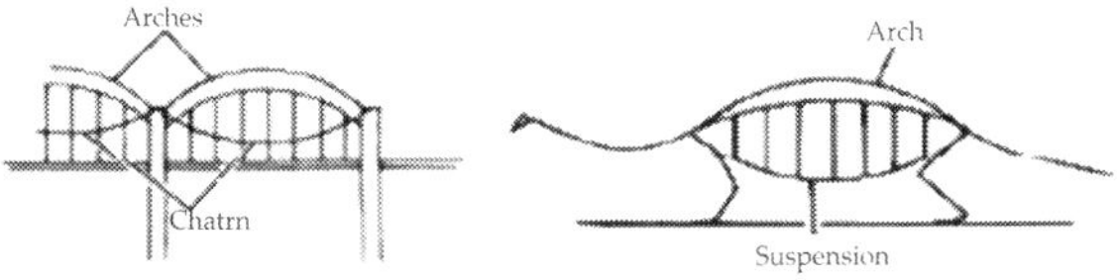

Royal Albert Bridge and Brontosaurus: (Images courtesy: Mehdi Sadri et al., 2014)

The Gateshead Millennium Bridge in England also is known as the 'Blinking Eye Bridge" by virtue of its tilt. It also won best architecture of the year in the country.

(Left)Blinking Eye Bridge: http://www.trekearth.com/gallery/Europe/United_Kingdom/England/Tyne_and_Wear/Gateshead/photo45337.htm. (Right): Prototype Courtesy: CHEN FEI and SHA SHA, 2005

It is a movable bridge and mechanism of opening and closing is very similar to that of characteristics of the human eyelid (CHEN FEI and SHA SHA, 2005).

Another iconic building that looks at bionics is the latest museum in Tianjin. The wings of a swan when spread has been used as a model for this construction and framework of the entire monument. Other constructions that have used bionics include Daidaimu Gymnasium in Japan based on a whirlpool and Beijing Olympic Assembly Room that was modelled on the nest of a bird (CHEN FEI and SHA SHA, 2005).

Several other instances of the use of bionics in the design of several buildings include: Flectofin TM is an application that uses Bird-of-Paradise-Flower or Strelizia reginae as its inspiration. This research study won the International Bionic Award 2012 and the Techtextil Innovation Prize in 2011. A prototype of a construction is being worked upon by ITKE-University Stuttgart, Albert-Ludwigs-University of Freiburg, ITV-Denkendorf, Clauss-Markisen.

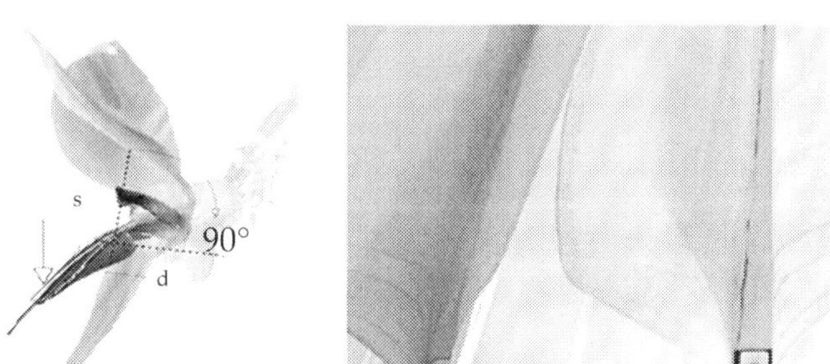

The Bird-of-Paradise-Flower and a prototype of Flectofin TM. Courtesy: Andreas Hammer, 2016

The use of thin-film photo voltaic modules in the form of flaps absorbs sunlight that leads to flaps made of several resins to curl. This curling exposes the photovoltaic- modules to the sunlight in a direct manner.

In a further advantage of this model, a team worked on the above system and added an energy collecting photo voltaic system. They used a clover leaf model that opens up for photosynthesis to absorb the sun rays for power collection by the flaps. With the use of nithinol-wire gaze in order to induce resistance upon being exposed to energy the flaps can be changed from open to close. This is aimed at minimizing overheating and the optimum use of solar energy. The use of shadowing devices to in the construction of buildings has been warranted. Several candidates from nature that show potential include snake-star- *Ophiocoma wendii* or a seal skin based system called Rete Mirabile.

Shadowing device from snake-star.

Courtesy: Andreas Hammer, 2016

Overall the use of biological principles in the construction of suitable buildings is showing promise. An interesting offshoot is the use of such flower or clover leaf mechanisms that open and close according to the time of the day, in order to make buildings more energy efficient (Andreas Hammer, 2016).

BIONICS AND BRIDGES

The Rainbow Bridge in China was constructed in southern Song dynasty more than 8 centuries ago. Inspired by a rainbow it was a mystery and was reconstructed using interlocking wooden planks subsequently.

Rainbow Bridge in China

Left: Cat's Cradle Bridge; Right: Shell inspired bridge

The winners of the best projects for bridge design in Foshan City in Guangdong Province, China were based on bionics.

One project used a principle of a cat's cradle that refers to a child's game while another looked at a design inspired from a shell.

MATERIALS USED IN CONSTRUCTION OF BIONIC STRUCTURES

Research from around the world looks at the potential of several materials in the construction of bionic bridges. They include:

- The principle of hollow interiors that imparts strength and endurance to many forces such as bamboo, bones of birds and reeds. Apart from being light weight, but they are resistant to bending and breaking.
- Queensland nut shell system that possesses fibers that extend in all directions encompassing the entire surface. The use of such a system can enable the construction of durable structures that can have stability and durability.
- A research team has added concrete to long fiber molecule

additives that has resulted in the formation of a net that is similar to a spider web. The main advantage in this system seems to be its potential in underwater constructions.

- A conch shell displays high resistance to stress, more than that of cement. Unlike the manufacture of several chemicals in construction that require heat, the product of nature requires no such interference.
- A water-soluble cement has been reported made of furfural-hydroxyketone polymers. It has displayed high amount of strength and also has an easy process of manufacturing (Reviewed by Chen Fei and Sha Sha, 2005).

CHALLENGES IN THE APPLICATIONS OF BIONICS IN BRIDGES

- A major challenge is the onslaught of natural elements such as wind, earthquake, rain, snow etc. The concept can be how to minimize the resistance to such elements that as the stronger the structure is, the greater its resistance is and the greater the counterforce of nature will be.
- Another challenge is to look at the effect of temperature on the integral linkers of a bridge. The increase and decrease of temperature causes expansion and contraction of the bridge materials resembling a human lung in action. Additionally, the mechanism can be likened to the co-operative movement of ligaments and skeleton of a person in locomotion. Though a human body is subjected to force when locomotion occurs, the elastic ligaments keep the bones in order. The application of such a principle to the concept of bridges can aid in the lessening of the impact of external forces on bridges.
- The use of colors and shape of a particular bridge are dependent on the environment where it is constructed. The use of mechanical construction in the construction fails to hide all the forces and damage a bridge is subjected to.
- This warrants the use of organic construction that has its roots in bionics. The process involves the building of a structure from

the interiors in a holistic fashion such that it is in sync with its surroundings rather than appearing as an addition (Reviewed by CHEN FEI and SHA SHA, 2005).

BIONICS IN THE DESIGN OF FORCEPS

The use of anatomical forceps is seen in the handling of delicate parts such as soft tissues, blood vessels etc. This use avoids the conventional instruments that could damage sensitive structures or lose their grip.

With a view to apply bionics in the design of such as forceps, a team in 2016 looked to the Fin-Rays of bony fishes for inspiration.

The Fin ray inspired the fin ray effect that is an isosceles triangle with elastic sides and stabilizers that deforms upon application of force rather than bending to a side.

Following the analysis of optimal height-width ration studies, the team used -aided design software SolidWorks 2015 Premium (Dassault Systèmes SolidWorks Corp.) to design models. This was followed by 3D-printing and subsequent manufacture and testing.

The benefits of the fin ray effect were mainly the adaptation of the jaws of the forceps to the object that it is used to handle. Apart from a uniform distribution of forces on the object that leads to decreased pressure on the material, the chance of slippage reported is also decreased.

The team reported the use of the bionic forceps in the handling of sensitive materials without damage to serve as an augmentation to the anatomical forceps used in medicine (Lucia Kölle and Oliver Schwarz, 2016).

BIONICS AND HARVEST OF WIND ENERGY

The use of wind for generation of power is being pitched as a renewable and clean technology (V. Fthenakis and H. C. Kim, 2009). The blades are the principal parts of the wind turbine used for this purpose. It has been reported that the blades are made up of several air foil surfaces whose aerodynamic features are vital to the output of the blades. The airfoils have been developed by the use of inverse designs. However, the use of bionics finds application in the design of such airfoils. Here, research points to using inspiration from the bird kingdom, sea gulls and Long-eared Owl wings (Xu et al., 2010).

Taking this notion ahead a team in 2016 designed a novel airfoil using a wing of the Long-eared Owl and tested in a wind turbine. The authors performed Numerical simulation analyses to evaluate the performances of standard blade and the bionic blade under typical wind conditions.

The authors reported an increase in the wind turbine with bionic blades as compared to the conventional wind turbine. It was also reported that the bionic wind turbine shows a power of 320W while the standard wind turbine is 265W at a wind speed of 7m/s, also suggestive of more efficient utilization of wind energy.

An enhancement of effective Angle of attack with an increase in the bending was displayed by the bionic blades. This resulted in a boost in the airfoil lift due to the difference in pressure between the upper and lower surface of the blade. Thus, the authors reported better aerodynamic features of the blade designed using bionics (Weijun Tian et al., 2017).

BIONICS AND PATENTS

A research team in 2011 proposed patent analysis as a suitable method opening various applications to unravel questions such as, "Which patents have its basis in bionics, and which inventions consist of bionic principles?" or "What are the characteristics of bionic-related patents?".

Following semantic patent analysis, the team searched for the word bionic in the US patent office (USPTO) database. Subsequently, extraction of first stage simple semantic structures using tools like semantic processor (Natural Language Processor (NLP) was performed. This was subjected to similarity analysis, matrix construction and positioning to represent 147 patents for the time span, 1977 to 2006.

It was reported that most of the patents belonged to the medical field and suggested that the work was a bridge between between the engineering-driven field of bionics with TRIZ and its different developments. The authors also suggested altering the usual slogan "learning from nature to inspire human innovation" to "Using patent analysis to identify and map bionic solutions extracted from nature" (Lothar Walter et al., 2011).

BIONICS IN THE DESIGN OF GANTRY MACHINES

The use of bionics was explored in the design of a large-scale gantry machining center. These machines were expected to have an increased stiffness along with vibration resistance, noise and movement stability. Previous work had explored the use of veins of waterlily leaves while this work was inspired by beehives. The use of Honeycomb core: triangle honeycomb, hexagonal honeycomb and combination of honeycombs called Kagome honeycomb was tested. Following analysis and comprehensive evaluation, it was reported that:

The column structure of ribs with "hexagonal honeycomb" shape has the best comprehensive performance in comparison with that of other shapes tested such as "X", triangle and kagome.

The use of honeycomb structures in the design of the ribs of the column of the machine led to a better performance, thus highlighting the use of bionics in the design of such machines (Yuhong Zhang et al., 2012)

SUMMARY

The applications of bionics seem to extend to non-medical fields as discussed in this chapter. The case studies examined included the use of bionics in architecture: buildings and bridges. Several buildings from the past have used bionics for their construction and so are many researchers currently using bionics for the construction of energy efficient buildings.

The field of bionics also has been applied in the construction of tools or machines such as the sensitive forceps in this chapter or the gantry machine. Another example was the use of bionics in the successful design of airfoils with improved aerodynamic conditions when a bionic approach was used as compared to conventional methods.

The field of bionics thus shows immense potential across several fields as seen in these chapters. Further research in the expansion of the various technical difficulties can augment the application of this novel biotechnological aspect into every aspect of life.

REFERENCES

1. David BP. The Tower and the Bridge: The New Art of Structural Engineering". New York: Basic Books, Inc., Publishers (1983).
2. Dongqing J (2003) Bionic building material and its applications. Building Material Development Guide.
3. Fei C, SHA S (2005) An introduction to bridge design based on bionics. Proceedings of the 24th Southern African Transport Conference (SATC 2005). 11 – 13 July 2005 ISBN Number: 1-920-01712-7. Pretoria, South Africa.
4. Fthenakis V, Kim H-C (2009) Land use and electricity generation: a life-cycle analysis. Renewable and Sustainable Energy Reviews 13(6-7), 1465–1474.
5. Giesecke S, Hafner-Zimmermann S (2003-2004) Bionics: Applying Nature's Principles for Intelligent Building. The European Foresight Monitoring Network. Foresight Brief No. 17.
6. Hammer A (2016) Climate Adaptive Building Shells for PlusEnergy-Buildings, Designed on Bionic Principles. World Academy of Science, Engineering and Technology. International Journal of Civil, Environmental, Structural, Construction and Architectural Engineering 10(2).
7. Kölle L, Schwarz O (2016) Bionic forceps for the handling of sensitive tissue. Current Directions in Biomedical Engineering 2(1), 91–93.
8. Podborschi V, Vaculenco M (2002) Study of natural forms: the source of inspiration in the product design, Republic of Moldova, Technical University of Moldova.
9. Sadri M, Kavandi M, Jozepiri A, Teimouri S, Abbasi F (2014) Bionic Architecture, Forms and Constructions. Research J Recent Sci 3(3), 93-98.
10. Salvadori M (1990) Why Buildings Stand up, W W Norton, London.
11. Tian W, Yang Z, Zhang Q, Wang J, Li M, Ma Y, Cong Q (2017) Bionic Design of Wind Turbine Blade Based onLong-Eared Owl's Airfoil. Hindawi Publishing Corporation. Appl Bionics Biomech Article ID 8504638. http://dx.doi. org/10.1155/2017/8504638.

12. Walter L, Isenmann R, Moehrle MG (2011) Bionics in patents – semantic-based analysis for the exploitation of bionic principles in patents. Procedia Engineering 9, 620–632.
13. Williams AH (2003) Zoomorphic: New Animal Architecture. New York: Harper Design International.
14. Xu C-Y, Qian Z-H, Liu Q-P, Sun S-M, Ren L-Q (2010) Aerodynamic performance of bionic coupled foils based onnleading edge of long-eared owl wing. Journal of Jilin University (Engineering and Technology Edition) 40(1), 108–112.
15. Zhang Y, Ye W, Leng S, Wenyong Shan W (2012) Bionic Design for Column of Gantry MachiningCenter Using Honeycomb Structures. Proceedings of the World Congress on Engineering Vol III. WCE 2012, July 4 - 6, 2012, London, UK.

CITATIONS

CHAPTER 3

Rouse EJ, Villagaray-Carski NC, Emerson RW, Herr HM (2015). Design and Testing of a Bionic Dancing Prosthesis. PLoS ONE, 10(8): e0135148. DOI:10.1371/journal.pone.0135148.

CHAPTER 5

Ayton LN, Blamey PJ, Guymer RH, Luu CD, Nayagam DAX, Sinclair NC, et al. (2014). First-in-Human Trial of a Novel Suprachoroidal Retinal Prosthesis. *PLoS ONE*, 9(12): e115239. DOI:10.1371/journal.pone.0115239.

CHAPTER 6

Malcolm P, Derave W, Galle S, De Clercq D (2013). A Simple Exoskeleton That Assists Plantarflexion Can Reduce the Metabolic Cost of Human Walking. *PLoS ONE*, 8(2): e56137. DOI:10.1371/journal.pone.0056137.

CHAPTER 8

Yiyun Zhu, Chang Jiang Wang, and Diane Mynors, "Applications and Engineering Analysis of Lotus Roots under External Water Pressure," *Applied Bionics and Biomechanics*, vol. 2016, Article ID 2386982, 8 pages, 2016. DOI:10.1155/2016/2386982.

CHAPTER 9

Ulrich Lüttge. Physics and the molecular revolution in plant biology: union needed for managing the future. *AIMS Biophysics*, 2016, 3(4): 501-521. DOI: 10.3934/biophy.2016.4.501.

CHAPTER 10

Mazzolai B, Beccai L, Mattoli V (2014). Plants as Model in Biomimetics and Biorobotics: New Perspectives. *Frontiers in Bioengineering and Biotechnology*. 2014; 2:2. DOI:10.3389/fbioe.2014.00002.

CHAPTER 11

McCloskey MA, Mosher CL, Henderson ER (2017). Wind Energy Conversion by Plant-Inspired Designs. *PLoS ONE*, 12(1): e0170022. DOI:10.1371/journal.pone.0170022.

INDEX